Evolution of Land and Life in Oman: An 800 Million Year Story

Mohammed Hilal Al Kindi

Evolution of Land and Life in Oman: An 800 Million Year Story

Springer

Mohammed Hilal Al Kindi
The Earth Sciences Consultancy Centre
Muscat
Oman

and

German University of Technology
Muscat
Oman

ISBN 978-3-319-60151-9 ISBN 978-3-319-60152-6 (eBook)
https://doi.org/10.1007/978-3-319-60152-6

Library of Congress Control Number: 2017960906

Printed on acid-free paper

This Springer imprint is published by the registered company Springer International Publishing AG part of Springer Nature.
The registered company address is: Gewerbestrasse 11, 6330 Cham, Switzerland

For all of those who contribute to the protection of Oman's geological heritage so that it will remain an attraction for researchers and a scientific resource for future generations, and for those who discovered or continue to discover parts of the story of land and life in Oman.

Mohammed Hilal Al Kindi
March 2018

Foreword

Oman has fascinating geology that is only partly known. In the years, we have been privileged to work in Oman and we have encountered many textbook examples of geology as well as other features or fossils which we did not understand the full significance of. By investigating further, often with the help of experts, some of these have turned into exciting new discoveries, new fossil species and new insights into the long and fascinating story of life. There are still many outcrops in Oman that have never been visited by a geologist, and around the next bend or over the next ridge, there could be something new, of national or international significance. This is one of the joys of geology, and finding fossils, crystals and minerals is often the start of an interest, a hobby, a career and a passion. Through much of the past 800 million years, Oman has lain, as today, on the edge of a continent and ocean. Currents within these oceans carried life from one region to another. As continents rifted or collided and oceans opened or closed, new arrangements came into place. Even before geologists understood why, it was known that some fossils were closely related, even though now found on different continents thousands of kilometers apart. Oman's climate has changed considerably through time, sometimes being cooler, and at other times being hotter or wetter. Sea levels have also varied over a few hundred metres higher and lower than today. There have been episodes of crisis, the major extinctions, when almost all animals and plants died out. The few that survived often flourished and diversified into new niches (e.g. birds being a survival from the dinosaurs). The fossil faunas and floras of Oman preserve evidence of these former oceans, ocean currents, re-arrangements of the continents and oceans, changes in climate and sea level, and the extinctions and recoveries. Yet the fossil record is only a selective glimpse of what lived —those that were preserved. We know something about animals and plants that had hard parts, like shells or skeletons, but often little about their soft parts or of entirely soft-bodied organisms. Then, there are trace fossils, the marks, footprints or casts made by animals or plants feeding, moving and living. In some locations and formations, these are abundant, but have been little studied. They can provide clues to the soft parts of organisms, how organisms fed or lived, and about the environments in which sedimentary rocks formed. Fossils are of economic importance too, as geologists use them to understand the ages and environments of rocks when finding and developing resources. The changing sequence of fossil spores and pollen, due to climate warming some 300 million years ago, allows geologists in the oil industry to unravel the complex reservoirs of the Al Khlata sequence, for example. The vugs of fossil rudist shells form some of the best reservoirs in the Shuaiba and Natih rocks. The 'Story of Land and Life Evolution in Oman' is an account of what is known today from the perspective of an enthusiastic Omani geologist. Hopefully, it will inspire students to take an interest in the land and fossil heritage of Oman.

Alan Heward
The CEO of the Earth Sciences Consultancy
Centre, part time lecturer in the German
University of Technology in Oman

Preface

Of all sections in the book, the acknowledgement section is the most difficult to write because I owe such gratitude to so many people who have contributed directly or indirectly to the production of this book. It will take another book to thank all of these people.

Upfront, I would like to thank Mr. Mahmoud Sabeel Al Belushi for his support to get this published in its final format. Mahmoud has accompanied me in many field trips during the preparation of this work. He has spent tremendous efforts and time to make sure this work gets produced. Thank you very much Mahmoud.

I would like to thank Jill Newby James for her great support in revising, correcting and occasionally rewriting part of the English text of this book. Beyond the language, Jill suggested important modifications in the book. Thanks also go to Dr. Alan Heward, Dr. Randall Penney, Suliman Al Furqani, Dr. Ibrahim Al Ismaili and Dr. Mutasim Al Gammari for their reviews. I would also like to thank Oscar Sanisidro for drawing the conclusion figures at the end of each chapter. Oscar represented the main elements of different time periods, as described in this book, with great skill. He also produced high-quality representations of the dinosaurs and the Cenozoic mammals. The book has benefited a great deal from discussions and geological trips with Prof. Ken Glennie, Dr. Juma Al Belushi, Dr. Mahmoud Al Mahrouqi and Hamad Al Wardi.

The geological surveys done in Al Khawd and Fanja, initially with the main discoverer of Cretaceous fossils, Dr. Axel Hartman, and later with a team formed by the Geological Society of Oman including Dr. Martin Pickford, Dr. Axel Hartman, Yousuf Al Sinani and Mohammed Al Waely, have revealed many new Cretaceous fossils, like dinosaur bones, crocodile teeth and turtle bones. The author also spent time in the Al Ghabah area in Central Oman with a group of scientists from the National History Museum of Paris, and the team found a number of mammals' fossils belonging to the Cenozoic. My sincere thanks also go to Jeffery Rose and his wife for joining us on our archaeological trip to Dhofar. Dr. Jeff kindly provided a lot of material on the Palaeolithic artifacts of Dhofar. I would also like to thank the late Prof. Maurizio Tosi from the Ministry of Culture and Heritage for providing archeological materials, including photos and articles, about the Neolithic and Bronze ages in Oman. I am also grateful to the heads of the earth science departments at Sultan Qaboos University and the German

University of Technology in Oman, Dr. Salah Al Khirbash and Dr. Michaela Bernecker, for their kind support.

I would like to specially thank Mahmoud Sabeel Al Belushi, for his kind support while visiting many outcrop locations across Oman. Many people have supported the author to know about many geological locations in Oman, including Fahad Al Thani, Majid Al Miqbali, Ahmed Qatan, Dr. Abdulrahman Al Harthi, Dr. Mohammed Al Wardi, Yousuf Al Shihi and Abu Nabil Al Shihi, Saleh Al Hikmani, Saif Al Rawahi, Heitham Al Amri.

The writing of this book has provided an excellent opportunity to meet and get work with many great Omanis and expatriates.

Muscat, Oman Mohammed Hilal Al Kindi
January 2018

About the Book

We have chosen to put the references at the end of the book to enhance the readability and maintain the flow of the text for specialists and laymen alike. The list at the end of the book contains most of the references used in this book.

The published fossils were found while filming *The Story of Life in Oman* or else the geologist who found the fossil is mentioned in the caption. In a few cases, the geologist is unknown. Most of the microscopic fossils have been found and classified by previous geologists from Petroleum Development Oman.

All the 3D graphic and animation works in this book are designed by the author for the films The Geological Heritage of Oman and The Story of Life in Oman. They have been executed by Stum Art Production, Oman. We have included a limited glossary, particularly for laymen, which may help to explain some of the terms in these pages without recourse to a dictionary.

Contents

About the Author

Mohammed Hilal Al Kindi obtained his bachelor's degree in physics and geology from the University of Aberdeen, Scotland, in 2003. He completed his Ph.D. with the University of Leeds, England, in 2006, studying the fractures and faults at the southern limit of the Al Hajar Mountains. In 2008, he became the executive manager of the Geological Society of Oman (GSO), and in 2013, he became the president of the society. Between 2006 and 2014, Mohammed worked as a petroleum development geologist in Petroleum Development Oman (PDO). He is a geological explorer, enthusiastic fossil hunter and keen photographer. His main specialism is structural geology and plate evolution. He has had the opportunity to visit many geological wonders and fossil-rich areas in Oman while working in PDO and during his contributions to GSO's projects. He has led a number of geological expeditions and research groups in Oman. Mohammed currently (2018) is the CEO and founder of the Earth Sciences' Consultancy Centre.

Introduction

This book will take us on a fascinating journey to discover the story of land and life in Oman from the very earliest evidence of 800 million years ago. How did the land of Oman evolve over millions of years? What organisms lived here? How did they develop? Did they become more complex and varied or become extinct and disappear forever?

Our journey will explore the movement of the land of Oman throughout the geological ages and explain how the land's vitality influenced the diversity of its organisms. We will move through time and fly through space to explore this dynamic corner of our planet. Together, through the scientific discoveries made in Oman, we will begin to understand the relationships—the harmony and conflicts—between the earth's environment and the organisms that live on it. Ours will be a journey of joy and discovery as Oman reveals its great secrets about the emergence of life on our planet. Its rocks record multiple changing environments, and its fossils represent the oldest evidence of multicell organisms ever found. Here lived the first primitive fish to swim in the sea and the earliest plants to grow on earth. Dinosaurs and elephants once roamed the plains. Early man passed through and even settled. Civilizations flourished in this land once called Majan and Mazoon. Welcoming and vibrant, amongst all this splendour, Omanis live today.

'The Story of Land and Life Evolution in Oman' is inspiring and could help the specialist and laymen to better understand similar stories in other parts of the world.

Afterword

While walking through the mountains and across the plains of Oman, it is very likely that you will step on or see an extraordinary fossil. You may as well wonder why do rocks have so many different colours. There is actually no surprise in that, because this land has hosted a myriad of life forms and environment changes holding many secrets about the origin and evolution of land and life over hundreds of millions of years.

Although there have been many immensely important geological discoveries in Oman, there remains a challenge to explorers to add their scientific footprints and discover new information about the wealth and diversity of the evolution of land and life in Oman since the formation of earth and begging of life.

1 An Outline of the History of the Earth and Oman

The summit of Jabal Shams on top of the Wilayat of Al Hamra, with its colorful surrounding rocks.

M. H. Al Kindi, *Evolution of Land and Life in Oman: An 800 Million Year Story*,
https://doi.org/10.1007/978-3-319-60152-6_1

1.1 Our Kinaesthetic Planet

Since its creation with all the other objects of the solar system some 4,600 million years ago, the earth has never stopped moving. Every day it spins on its axis at over 1,600 km/h along the equator. Every year, it also rotates around the sun, travelling at a speed of approximately 110,000 km/h. Beyond these dizzying movements, as an object in the solar system, the earth also travels around the Milky Way once in every 240 million years. Relative to the galactic centre, it reaches an astonishing 830,000 km/h. The galaxies, with all of their visible and invisible—or dark—matter, also spin around themselves, rotating rapidly amongst a frame of galaxies in the vast universe. It is because of these rotations that the outward, centrifugal forces balance the gravitational inward pull and keeps everything more or less in place.

In addition to the movement within the system of the universe, everything on the surface of the earth is moving and changing, too. The continents pass from one place on the earth's surface to another through geological ages at rates reaching up to tens of centimetres per year. As this movement happens, mountains arise in the collision zones between continents and gulfs where the continents are pulled apart. Mountains, however, do not remain the same. They are slowly demolished—eroded by wind and rain that sculpture their faces and rivers that punctuate their cliffs. This ceaseless process of construction and demolition and the endless creep of continents from one place to another generate new climates and environments, extinguishing some species and creating others in an infinite process often incomprehensible during the short and limited lifespan of humans.

The wonderful kinaesthetic nature of the earth is a direct product of its wonderful physical, chemical and biological composition. Humans have in the last centuries spent tremendous efforts to understand the various global processes that control the evolution of this planet and life on it. They have acquired substantial knowledge, but there is still a lot to be discovered. For instance, humans are still on the process of identifying all the natural variables that make this planet hope for so many life species and whether there is any tiny possibility that these many variables may exist in any other planet in the universe (Fig. 1.1).

Fig. 1.1 Earth

1.2 The Earth's Components

The earth was most likely born by the process of accretion in a molten state. As its surface cooled, heavy elements such as iron and nickel sank to its heart, while the lighter elements such as silica floated to the top. This process led to segregation between the elements deposited in the interior of the earth and those at its surface. In fact, there are three layers with differing chemical compositions, namely the core, the mantle and the crust (Fig. 1.2). The core consists of iron and a small percentage of nickel and forms the densest part of the earth. The core is divided into internal and external parts. The temperature of the inner core exceeds 5,500 °C, a temperature almost equal to that of the surface of the sun. In spite of this immense heat, the materials of the inner core exist in a solid state because they are under extreme pressure. The outer core, on the other hand, exists as a low-viscosity liquid. This is because of the equivalence between its pressure

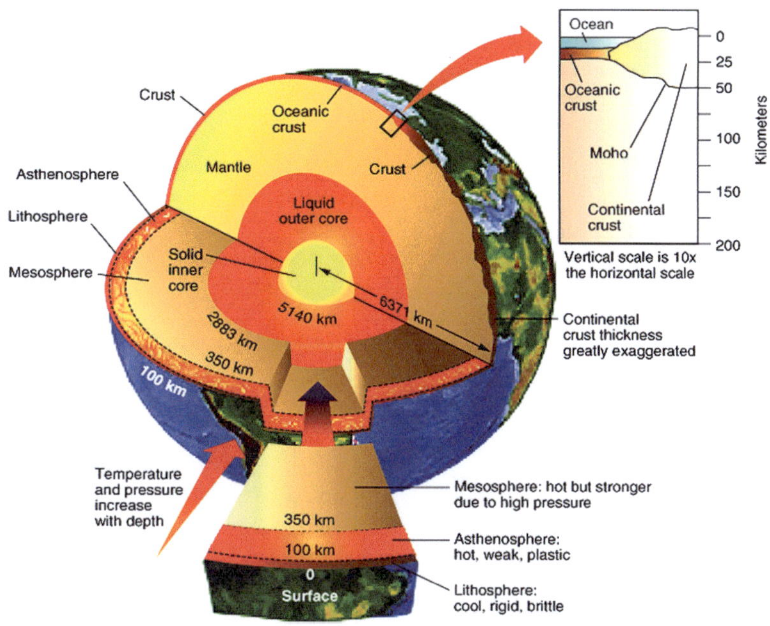

Fig. 1.2 Chemical layers of the earth (core, mantle and crust) and the physical layers (lithosphere, asthenosphere, mesospheric mantle, outer core and inner core), after Skinner and Porter, 1987 (Published by John Wiley and Sons)

and temperature since the pressure at the surface is much lower than at the core. If cracked open, the earth would disgorge its hot, heavy and viscous internal fluids, consuming the light, thin, covering crust and causing the earth to return to its original form as a ball of flame. Happily, this state has never been experienced nor could it be comprehended by humans.

The earth's second chemical layer is known as the mantle. This layer represents around 85% of the earth's size, and it is cooler and less dense compared to the core. Although predominantly solid, the materials of the mantle behave like a very viscous fluid over geological time and move by convection cycles at rates of up to 20 mm a year near the mantle's top.

The earth's outer skin is called the crust. The line separating the crust from the mantle is known as the Moho, after a Russian scientist named Mohorovičid. It can be seen exposed in a number of places in Oman because of previous tectonic plate action. The most famous site is at Wadi Al Abyad in North Oman (Fig. 1.3). The crust covers the earth's molten interior with a—relatively—very thin layer of cold, solid rock—sedimentary, metamorphic and igneous. There are two types: the oceanic crust which is typically less than ten kilometres thick and the continental crust which averages around thirty-five. The radius ratio of earth's crust, which we live on, compared to the radius of the whole earth is only 0.5%. However, without this thin mattress on the top of a solid foundation of warm and dense material, life—including ours—would have been unsupportable on an otherwise very hot earth.

In addition to the chemical subdivision of the earth's layers, the earth is also physically or mechanically divided into five layers based on the rheological state of its matter from solid to liquid. This division includes the lithosphere, asthenosphere, mesospheric mantle, outer core and inner core. The lithosphere is the hard, rigid outer layer of the earth, which includes the crust and the upper mantle. It is divided into two plates, known as oceanic and continental lithospheric plates. The asthenosphere is the viscous layer of the earth. It comprises the area between the upper mantle and the lithospheric mantle and in some regions reaches depths of 700 km. The asthenosphere is slowly flowing at a rate of few millimetres per year as a result of the convection heat cycles that originate from the earth's interior and circulates outwards towards the crust.

Fig. 1.3 Line of Moho that separates between the mantle and the crust of the earth can be seen in Wadi Al Abyad in the Wilayat of Al Maawil (where the red line is). The dark rocks of the mantle appear below the line, whereas the light coloured rocks of the crust can be seen above

1.3 The Age of the Earth

The age of the earth can be represented as a clock. Figure 1.4 illustrates the main time periods of the earth's history since its creation. The Palaeozoic, Mesozoic and Cenozoic Eras are known together as the Phanerozoic eon. The boundary between the Precambrian and the Phanerozoic (541 million years ago) has been determined from the subsurface rocks of the South Oman Salt Basin.

The age of the earth, which spans a period of 4.6 billion years, is divided into four main eons. The first three are the Hadean, Archean and Proterozoic. These eons are informally known as the Precambrian time or the Cryptozoic Supereon. The fourth eon is the Phanerozoic, and it is composed of three eras: the era of ancient life (Palaeozoic), the era of middle life (Mesozoic) and the era of new life (Cenozoic). The latter is the era in which we live today. Had we made the whole age of the earth as one clock cycle of 12 h (Fig. 1.4), then the Precambrian—the very ancient history of the earth—will represent ten hours of time. It is a time we know very little about, although this is the period when life on the earth began. During the following three eras (Palaeozoic, Mesozoic and Cenozoic) of the Phanerozoic, life diversified enormously as evidenced by the amount of fossils visible in the layers of rock deposited during this eon—hence, the name Phanerozoic which means 'visible'. It is worth mentioning that the boundary between the Precambrian time and the Phanerozoic Eon was first defined in Oman from the existence of small organisms known as Cloudina and Namacalathus which became extinct between the two periods. The layers that mark this extinction are separated by a lithostratigraphic unit dated to 541 million years ago. This age is now universally adopted to define the end of the Precambrian time and the beginning of the Phanerozoic eon.

Life first began in the earth's oceans more than 3.5 billion years ago—less than one billion years after the earth's creation. It took the form of prokaryotic bacteria: single-celled organisms with no well-defined nucleus. These frequently form layered domal shapes known as stromatolites. Simple multi-cell creatures—like sponges—only appeared some 630 million years ago. About 100 million years later, the arthropods, fish and other marine organisms began to flourish. Around 450 million years ago, the first plants emerged on land, followed by insects and amphibians 400 and 350 million years ago, respectively. By 300 million years ago, reptiles had occupied the land and begun spreading. Mammals appeared about 200 million years ago, but only dominated by the extinction of dinosaurs 65 million years ago. The first birds appeared around 150 million years ago, and flowering plants blossomed 130 million years

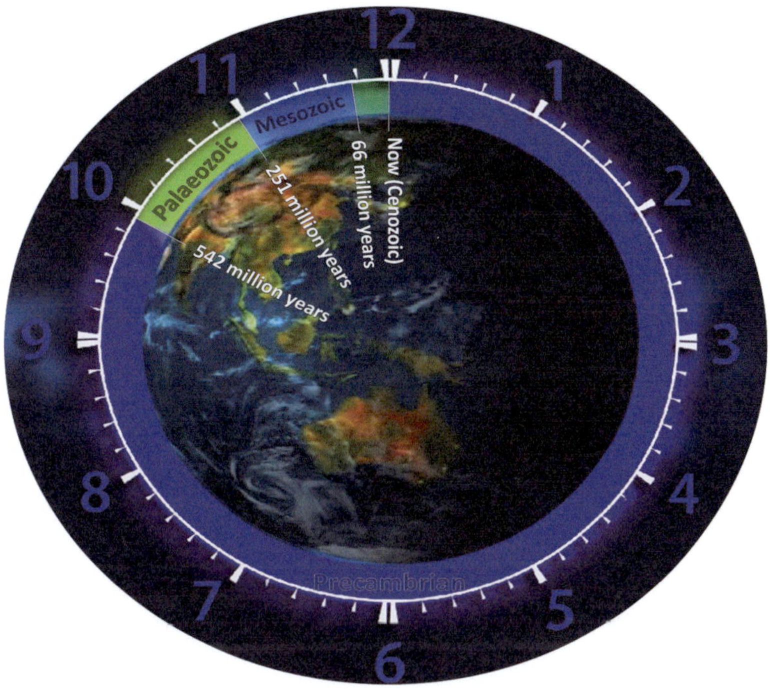

Fig. 1.4 Earth's clock represents the eras of time and their correlation to one rotation of the clock

ago. In contrast, humans have only been here for the last 200 thousand years and have witnessed a mere 0.0043% of the earth's history.

Life reached its present form as the last of four distinctive phases in the earth's evolution. In the first phase, around five billion years ago, the earth became a separate and rounded planet within our solar system. Super-hot dust and gas from an exploding star began to clump together. This accretion of solar nebula formed the sun and a number of orbiting planets: the earth was the third farthest from the sun. In the second phase, the earth began to cool, allowing the heavy elements to descend to the core and the lighter, colder elements to form the crust of mountains and plains. The third phase, around four billion years ago, was the formation of an atmosphere and oceans. Here, earth's position was fortunate —any closer to the sun and the water would have boiled and evaporated, and any farther away and it would have frozen. The stage was now set for the final phase or phases: the start and evolution of life on earth. The water, atmosphere and proximity to the sun distinguished earth from all the other planets in the solar system. After hundreds of millions of cycles of cooling and heating, the earth became the perfect environment for complex life forms to develop and flourish.

1.4 Tectonic Movements

The earth's interior is in constant motion. The difference in temperature between the core and crust generates heat currents that circulate upward from the core. As they move outward, they cool and circulate back towards the core in an endless convection cycle. At the surface, these currents are moving at an average rate of more than 4 mm per year, in a process known as plate tectonics (from the Greek tekton meaning 'builder').

As the currents ascend, they push magma to the surface of the earth along major cracks known as mid-oceanic ridges. This process is responsible for building new crusts and for causing volcanoes and earthquakes. Long, high mountain chains lie in the deep oceans along the mid-oceanic ridges. Some are more than 10 km high from base to top—taller than the Mount Everest. Newly formed crusts in the middle of an ocean are often compensated for by subsiding crusts on the sides of the plate. This process of crust recycling plays a vital role in the evolution and diversity of life. It is also significant in releasing excessive heat from deep inside the earth, protecting life on its surface.

Earth would have looked totally different if the mid-oceanic ridges were not covered by seawater. Had the emissions from these vents and cracks been directly released into the atmosphere, they would have eliminated life on earth.

The process of plate tectonics is responsible for the shift of the earth's continents from one place to another over millions of years. The surface of the earth is made up of at least fourteen plates. Most plates represent a suture of many micro-plates that accreted millions or billions of years ago. Among these plates is the Arabian plate, of which Oman forms the eastern edge.

Over the last 600 million years, Oman and Arabia have moved from a position near the equator to the South Pole and back to their present location, north of the equator (Fig. 1.5). As a result, Oman has experienced different climates and conditions over time. Various types of rock have been deposited, and numerous forms of life have emerged during this long journey. The layers of rock in Oman are an amazing record of the evolution of life during the operation of tectonic action. The fact that Oman has always been positioned on the edge of Arabia has led to a rich history of plate movement and contributed to the multitude of life forms to live on this land. Today, Oman is still moving slowly toward the northeast at a rate of about 2 cm per year. Overall, the Arabian Plate is surrounded by at least four major crustal cracks that include to the west the Red Sea spreading zone, to the south the Own Fracture Zone in the Indian Ocean, to the north the Zagros collisional zone between Arabia and Eurasia and to the north the Levant fault system. Therefore, the margins of this plate are currently active.

Plate tectonics have played an essential part in creating the diversity of life throughout earth's history. Separated and moving continental plates are like navigating ships carrying on their surface passengers or organisms that evolve differently in various environments and adapt according to the climatic conditions of their new positions on the globe. When continental plates collide or become connected through islands or narrow strips of land, new species of organisms may evolve and others become displaced as a result of the interactions between the two meeting groups of organisms.

The dynamic process of plate tectonics, therefore, not only continuously transforms the earth's surface, but also stimulates various patterns of evolution. Moreover, through the convection of the earth's molten interior, plate tectonics contribute to the generation and strengthening of the earth's magnetic field. This deflects the harmful solar winds which would erode the earth's atmosphere and thus protects life on earth. In addition, plate tectonics play an important role in the recycling of carbon and the generation of CO2 and other greenhouse gases through volcanic activity.

1.5 Determining the Absolute Age of Rocks

Radioactive elements are often used to determine the age of rocks and their fossil constituents (Fig. 1.6). Among these elements are uranium, potassium, rubidium and carbon. Uranium is commonly used to measure the age of rocks older than 100 million years, because it turns into different isotopes of lead in a half-life between 4,500 million years and about 710 million years. The decay of potassium to argon (K-Ar), in a half-life of up to 1,300 million years, is used to measure the age of rocks ranging from 0.1 to 3,000 million years. Rubidium is often used to date rocks older than three million years. The long half-life of rubidium makes it useful in determining the age of meteorites and the oldest rocks on earth. On the other hand, carbon is found useful to determine the age of rocks and fossils that are less than 70,000 years old. It is considered an excellent method of dating the

Fig. 1.5 Location of Oman (red dot) over millions of years, from number "1" at about 650 million years ago to number "21" that represents the land and Oman's position today. The first row on the left spans the period from 650 to 260 million years ago, the second row from 240 to 105 million years ago and finally the last row on the right is from 90 million years ago up to date. Light blue and dark blue indicate shallow seas and deep oceans, respectively. Note that during many geological periods, Oman was immersed in a shallow sea. Image (C) by Ron Blakey, NAU Geology

remains of human civilizations or recently deposited rocks. Among newly discovered dating methods is optical dating or the optically stimulated luminescence method (OSL). This technique measures the period of time since the last exposure of a mineral to daylight. It is

Radioactive element			Half life of parent (years)	Useful Range (years)
Parent		Daughter		
Uranium	Uranium-238	Lead-206	4.5 billion	10 million to 4.6 billion years
	Uranium-235	Lead-207	710 million	
	Thorium-232	Lead-208	13.9 billion	
Potassium-40		Argon-40	1.3 billion	100,000 to 4.5 billion
Rubidium-87		Strontium-87	47 billion	10 million to 4.5 billion years
Carbon-14		Nitrogen-14	5,700	100-30,000 years
OSL				300 to 100,000 years

Fig. 1.6 Methods used to measure the absolute age or the chronometric age of rocks and the time range of each method. For the young artefacts of humans, archaeologists use life properties to estimate the age

practical in determining the age of minerals typically deposited from 300 to 100,000 years ago, particularly in sand dunes and recent water deposits.

As for sedimentary rocks, which cover about 75% of the earth's surface but lack radioactive isotopes, relative dating is used. This method depends on stratigraphy—the sequential order of strata or layers of rock—in determining their relative age. Older layers of rock usually lie below younger ones; hence, their sequence reflects their relative age, with the oldest rocks at the bottom and youngest ones on top. Correlation of stratigraphic layers from one place to another can reveal the relative dates of fossils and rocks across large areas. Fossils of short-lived species can also be used as an index to identify the age of sediments.

1.6 What Are Fossils?

Fossils are the signatures of organisms whose traces are preserved in the rock layers. In fact, this term also includes non-organic marks, such as those made by drops of rain or the movement of waves. Scientists have so far identified more than 250,000 kinds of fossil from different times, and although that seems a lot, it is thought to represent no more than 10% of the actual number of organisms that have ever lived on earth. There are millions of animals and plants that have never been preserved in rock, particularly those that lived on land, since they could decompose on the surface before being buried in sediment. Moreover, organisms without a rigid shell or skeleton—such as insects—are often not preserved unless completely enclosed in a protecting envelope like amber.

Understanding the long-term geological processes and climate changes that promote biological diversity was almost impossible in the past. However, some early scientists had astonishing insights into the natural processes that result in the diversity, evolution or extinction of animals.

Among these scientists was the well-known Arab poet and philosopher Abu Othman Amr ibn Bahr Al-Kinani Al-Basri (notably known as Al-Jahiz), who wrote in the ninth century, some thousand years before the appearance of the Origin of Species by Charles Darwin. In his encyclopaedic book Al-Hayawan (Book of Animals), he made the following statement:

> Animals engage in a struggle for existence; for resources, to avoid being eaten by other animals or to breed. Environmental factors could influence organisms to develop new characteristics so that they can survive, thus transforming into new

> species. The organisms that manage to survive would be able to breed and therefore pass their special characteristics to their offspring.

Al-Jahiz was therefore able to predict and describe complicated evolutionary processes, such as natural selection and survival of the fittest, which explained the diversity of living organisms.

For thousands of years, humans must have observed the widespread phenomenon of seashells and marine snails on top of mountains and across plains and deserts far away from the sea. They have often mistakenly attributed this to the flood of the Prophet, Noah. They could not imagine, with their limited lifespan, tools and knowledge, the long, slow processes of nature and changes of environment that cause complete transformations of the earth's face over thousands and millions of years. Many Greek and Arab philosophers, however, provided reasonable arguments as they attempted to understand the existence of fossils in remote areas or peculiar places. The great Arab philosopher Ibn Sina in the tenth century said:

> Possibly they formed from sticky clay that dried up a long time ago and became rocks in unrecognizable times. It is likely that this inhabited place may have been not inhabited (by humans), but immersed in the sea and therefore solidified.

Ibn Sina clearly understood environmental change and his notion that living creatures became rocks is a good explanation of the process of fossil formation.

In fact, there are a number of ways in which fossils form, and what Ibn Sina would have seen would have been one of four main categories of fossils, illustrated in Figs. 1.7, 1.8, 1.9 and 1.10. They are moulds that represent a reverse impression in the rocks made by the organism, casts which form when the mould is filled by sediments or minerals, true form fossils where the actual animal bone or tooth is preserved and trace fossils which capture the activity of the animal, such as its burrow, footprints or gastroliths.

The process of preserving dead organisms varies significantly. Shells of invertebrates and the bones and teeth of vertebrates are often preserved because of their hard-resistive components. On land, most organisms decompose when exposed to the air. When deprived of oxygen in some way, their remains are saved. In rare instances, animals and plants are completely preserved without any significant change either by being entombed in tar or amber or by being encased in ice, like the mammoths of Siberia.

Another method is desiccation, where all moisture is drawn out of the organism in extremely arid conditions, preserving it whole. The most familiar means of fossilization, however, are carbonization and permineralization. During carbonization, soft tissue—in leaves or marine invertebrates, for example—decays, leaving only a carbon imprint on the rock on which they came to rest. Permineralization is the commonest form of fossil preservation and occurs when minerals such as silica fill the cells of the organism and crystallize. Sometimes all the carbon dissolves away, and only the shape and form of the creature remains, set in stone and trapped by later sediments.

Fossils are extracted from rocks using different methods. Phosphatic conodonts are extracted using acetic acid, fossils that are silicified are extracted using hydrochloric acid and chitinous fossils like acritarchs, chitinozoans, spores and pollen are extracted from shales using hydrofluoric acid. The fossilization process depends on the materials that make up the skeletons. For example, mollusc shells (e.g. snails and clams) and recent corals that are made of aragonite often occur as fossil casts because aragonite is unstable over geological time. Calcite skeletons are more resistant to weathering and they make up the skeletons of many animals, such as extinct corals (rugose and tabulate), brachiopods and echinoderms. Apatite (calcium triphosphate) is the main component of fossil bones and teeth of vertebrates.

1.7 Extinctions

Extinction is the death and complete disappearance of an entire type of organism. It mostly happens gradually and for a number of reasons.

Fig. 1.7 Cast fossils form when the actual mould of the fossil is filled in by another mineral, like calcite. This natural cast of a Mesozoic gastropod is about 140 million years old, and it was found on Saiq Plateau, Al Jabal Al Akhdar. The fossil is about 10 cm long

Fig. 1.8 Mould fossil: 450-million-year-old trilobite from Wilayat Al Amrat. The fossil is about 3 cm long

Fig. 1.9 True form fossil: dinosaur vertebra from the Al Khawd conglomerate. The fossil is about 15 cm long

Fig. 1.10 Trace fossils from Musandam, in North Oman, showing burrows of marine animals in limestone beds that were deposited in the Mesozoic Era in a shallow sea environment

Many years of continuing change go by before the last member of a species passes away. However, at least five mass extinctions have been identified during the history of the earth (Fig. 1.11). These are characterized by a significant decrease in the diversity and abundance of the observable life forms on earth. Because of their long and diverse fossil records through the geological timescale, marine life is often used as the index for the mass extinction events. Two of the five mass extinction events occurred during the Palaeozoic, 450 and 370 million years ago. During the first, about 60% of marine species disappeared (life had barely begun on land then) and in the second more than 50% of all species vanished, including plants and insects, which had already colonized the land. These two Early Palaeozoic mass extinctions can be considered together as the second largest identified extinction that affected the percentage of life genera on earth.

The worst ever life extinction is the great mass extinction between the Palaeozoic and the Mesozoic, about 252 million years ago, which destroyed more than 90% of marine species and up to 70% of land species. It took more than 10 million years for life to bloom again on earth after this extinction. Evidence of it can be seen in the rocks of Al Jabal Al Akhdar in Oman (see Fig. 58). This mass extinction lasted for about 200,000 years; therefore, the physical and biological changes lasted for a longer time compared to other extinction episodes and caused contemporaneous extinction on both marine and land species.

During the Mesozoic, two other mass extinctions occurred. The first happened about 200 million years ago and caused the death of around 50% of all species. However, the most famous extinction of all is probably the second Mesozoic extinction, some 65 million years ago, which wiped out the dinosaurs—along with hundreds of other species of plants and animals.

It is difficult to say exactly what the causes of extinctions are. Most of the related assumptions lack solid scientific evidence. Causes, however, could be abrupt such as meteoritic impacts. For instance, the mass extinction that occurred between the Mesozoic and the Cenozoic, which killed all dinosaurs, was explained by the impact of the Chicxulub Meteorite in the Gulf of Mexico. Causes could equally well be gradual, and

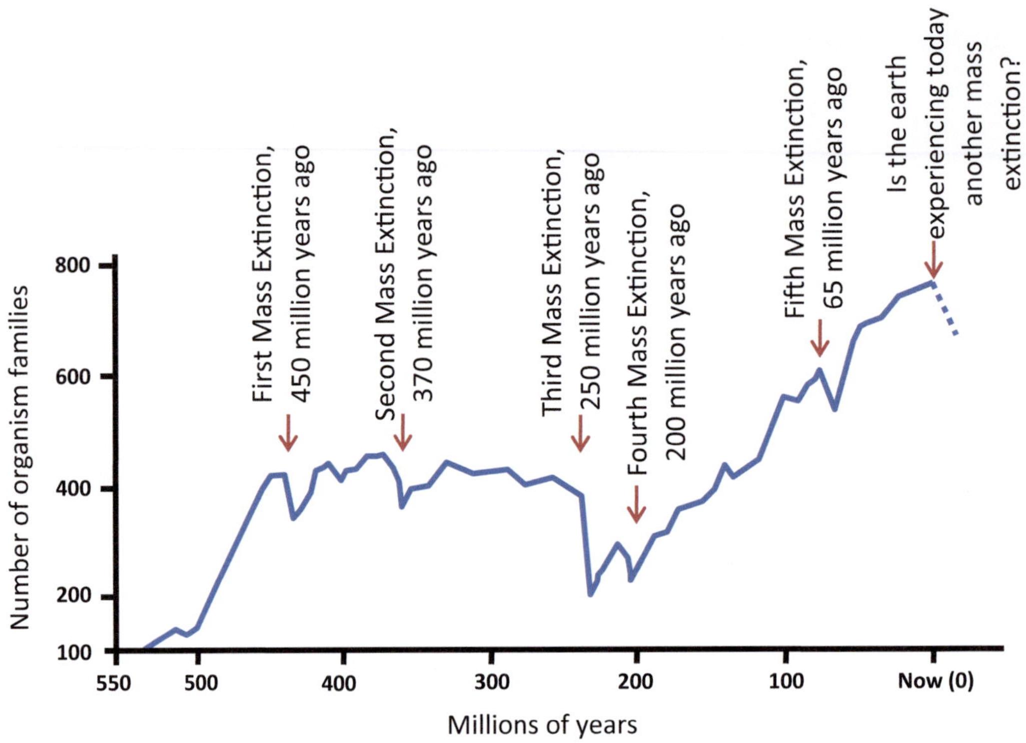

Fig. 1.11 Major periods of extinction during earth's history and their impact on the number of organism families (after Raup and Sepkoski, 1982)

these are often linked to global warming caused by increased volcanic activities, changes in the composition or condition of the atmosphere, or too low salinity and the consequent reduction in food sources in the oceans. One of these causes might have resulted in the greatest mass extinction at the end of the Palaeozoic.

At least 99% of all the organisms that have ever lived on earth have become extinct and no longer share life with us. Scientists estimate the number of species that have existed since the beginning of life on earth to be around 30 billion. About 10 million species still live today. Life on the earth has suffered from catastrophes in mysterious conditions, either during periods of mass extinctions or during other minor extinction events. This cyclicity of life was essential though, since through it we can see such diversity of life today. The dinosaurs would not have appeared on the earth without the extinction of many huge Palaeozoic reptiles and amphibians. Mammals only spread on the earth and became bigger and varied after the extinction of the dinosaurs and many invertebrates at the end of the Mesozoic. It appears then that extinction is actually a requirement for life to continue and that the death of some is fundamental to the survival of others.

The changes in the geological and ecological systems on earth can lead to almost complete extinction of all species on land and sea. They can also wipe out long-lived lineages and help to support the spread of less common and younger species. After mass extinctions, the forms of life often change and new ecosystems are established, where newly evolved species dominate.

However, since life on earth began, the number of families of an organism is—overall—continuously increasing, despite the occurrence of periodic extinction pulses. This suggests that mass extinctions are essential in order to create more hospitable environments for different organisms and support the evolution of new taxonomies that are probably less susceptible to environmental changes. Many scientists suggest that we are currently on the brink of a sixth mass extinction, directly or indirectly caused by humans. In the next few hundred or thousand years, this new event might lead to the extinction of nearly 90% of all species living today. It will differ, however, from all previous extinction events because the main cause is one of the species at risk—humans.

1.8 Classification of Organisms

Given the huge number of organisms that have lived on earth, biologists needed to introduce a classification which characterized them in a hierarchy based on shared similarities inherited from common ancestors. From the top, this order starts by kingdoms (e.g. Animalia, Plantae, Fungi and Bacteria), followed by phylum, class, order, family, genus and finally, species (Fig. 1.12). Humans, for example, are part of the Animalia kingdom, the Chordate phylum, the Mammalia class and Primates order. We could imagine this biological series like a large tree. The tree has a complex system of branches, yet it does not have a large trunk because the organisms have varied and branched very early in life's history. This tree continues to grow as new species branches are added to it every day.

Despite the hostile environment—lack of oxygen, searing temperatures, cosmic and solar radiation—life started on earth more than 3.5 billion years ago. The emergence of life is one of the greatest debates in the scientific world. How it happened has not yet been resolved and probably never will be. Many civilizations have offered explanations and theories, yet none of these theories has been proved adequate to recreate life. The basic elements needed for the formation of organic compounds such as carbohydrates, proteins and nucleic acids were available dissolved in water since those ancient days. The potential for life was like a soup with the right ingredients awaiting the torch to bring it to the boil. Some claim that the torch was a bolt of lightning or a meteorite strike, but these explanations have no verified scientific evidence.

The French chemist, Louis Pasteur, proved in the nineteenth century that the theory of spontaneous generation or anomalous generation proposed by the ancient philosophers from the time of Aristotle and beyond cannot be correct. The confined system remains lifeless unless a form of life is induced to it. This new understanding opposed the previous argument made by Greek philosophers who claimed that life could evolve from inanimate objects. One of the common examples used by the Greeks was the appearance of maggots in dead flesh after a period of time. In spite of the many theories that discuss the evolution and origin of life, none of them so far provide a convincing explanation. Even the well-known biologist Charles Darwin did not thoroughly discuss the emergence of the very first cell.

Among the nucleic acids essential to the emergence of life are RNA and DNA. These two acids have a very delicate structure and form in the nuclei of cells. It is important to note that even the prokaryotic cells, which represent the first form of life on earth and do not have a defined nucleus, contain what is known as nucleic acids. RNA plays an essential role in regulating and conveying genetic information to several proteins, whereas DNA encodes all the genetic information required for the development and functioning of all forms of life.

Fig. 1.12 Biologists classify organisms in a hierarchical order of seven sections from kingdom, which combines organisms with major similarities to species that share minute biological details

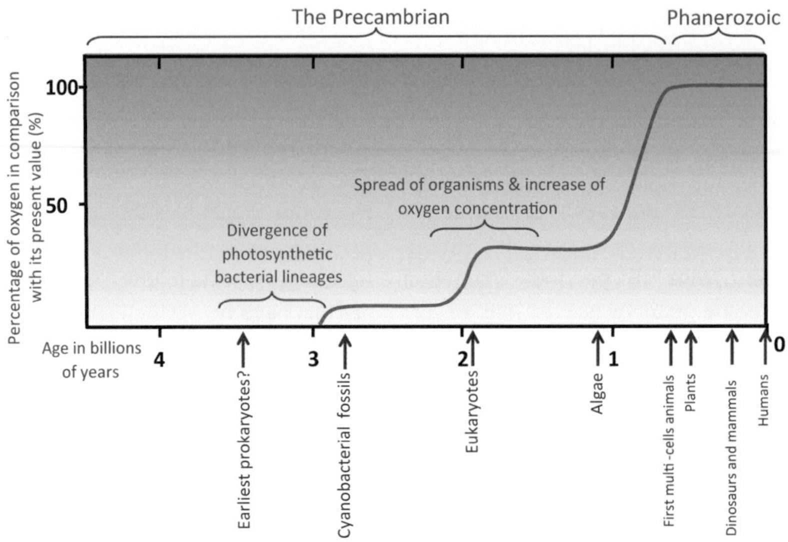

Fig. 1.13 Percentage of oxygen from 4.6 billion years ago and its effect on the pre-eminence and diversity of life, after Xiong and Bauer in 2002

Water is the most important element in the creation of organisms, much more important and directly related to the evolution of life than the concentration of oxygen in the atmosphere. In addition to being an original constituent of all living cells, serving as solvent, transporter, temperature regulator and lubricator of joints, it was also the system where life originated and evolved for more than three billion years before terrestrial organisms crawled out of the swamps and headed for land.

The liquid state of water distinguishes our planet from all others and covers about 70% of earth's surface—representing the most common chemical compound in the world. Nevertheless, it is not yet clear why there is more water on earth compared to many other planets. Did it form during the cooling down of the atmosphere when the pressure became optimal to hold hydrogen and oxygen atoms and cause a reaction between them? Or was it brought to earth by the comets and meteorites that peppered the surface after its formation? Has it been formed in large quantities only when oxygen was emitted by photosynthesizing bacteria? Evidence from rocks from the earth's earliest days suggests that water existed on earth shortly after its creation and that it fell in large quantities to form oceans, seas and rivers. These become both host and component of the very first organisms.

Another important factor in the evolution and diversity of life during certain periods of earth's history is the increase in concentrations of oxygen at particular times. Figure 1.13 illustrates the relationship between increased oxygen and the presence of organisms that perform photosynthesis like bacteria and algae. The blossoming of life at the end of the Precambrian and the beginning of Palaeozoic corresponds to leaps in the percentage of oxygen.

1.9 Major Geological Domains and Topography of Oman

Located on the eastern margin of the Arabian Peninsula, Oman has a wide spectrum of geological windows and terrains that capture the history of evolution of the Arabian platform and contribute significantly in understanding key earth dynamic processes (Fig. 1.14). This is primarily because Oman has a wide spectrum of rock formations that span significant windows of the earth history. These formations are well exposed in many places across Oman, and they have also been penetrated by many petroleum and mining boreholes that were drilled in different parts of the country at various depths, reaching several kilometres in some places.

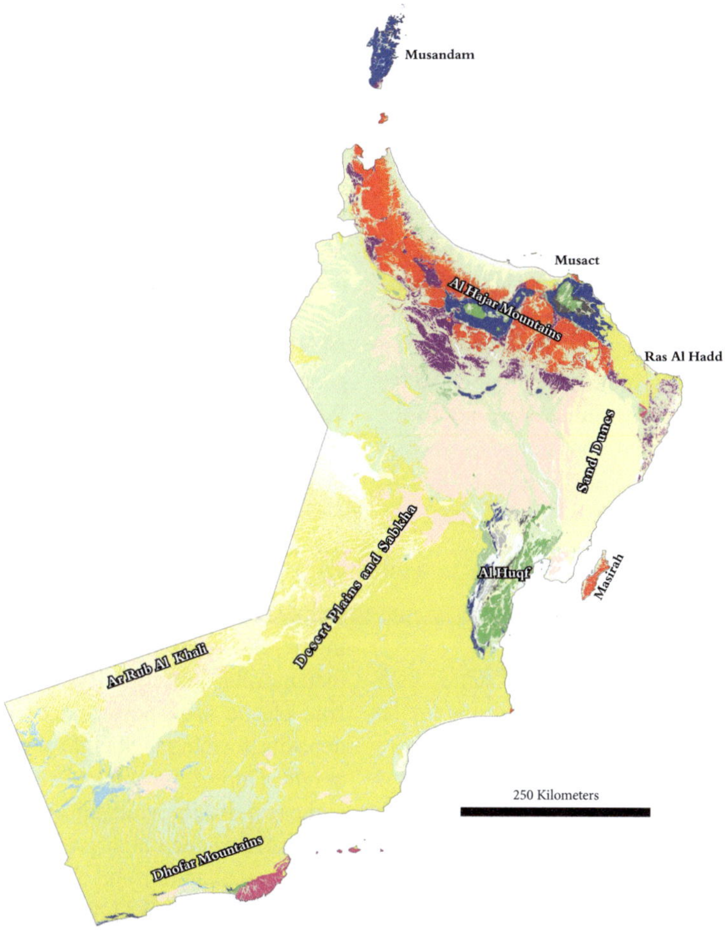

Fig. 1.14 A geological map of Oman; in summary, the warm colours (red and orange) in the north and in Masirah Island are for igneous oceanic a(Ophiolite), purple for emplaced oceanic sediments, dark pink in South Oman for Precambrian igneous basement rocks, dark green for Precambrian sediments (mainly carbonate and clastic rocks), black for Palaeozoic deposits, blue for Mesozoic formations and light colours (including yellow, blue and red) for Cenozoic deposits

The land of Oman, as the rest of Arabia, is chiefly covered by Cenozoic deposits, including early Cenozoic carbonate plains and late Cenozoic alluvial fans, as well as bodies of sand dunes, particularly exposed in central and South Oman.

The Sultanate of Oman covers an area of about 310 thousand square kilometres. Its coastal lines stretch for more than 3 thousand kilometres. Largely, Oman could be divided into five terrains, the Al Hajar Mountains in northern Oman with its associated alluvial fans, the Al Huqf Mountains and Masirah Ophiolite in Central Oman, the chain of the Dhofar Mountains in southern Oman with its alluvial plains, the desert and sabkha plains in central and South Oman and the sand dunes. Each morphological terrain has its specific geological record and wonders. They also differ significantly in their climatic setting throughout the year, therefore leading to significant diversity in environment and flora and fauna. Each terrain has its specific tourism destinations and other economic potentials. Geologically and archaeologically, the terrains also offer different attractions. Past and present, the communities living in these different terrains had various professions, customs and crafts.

The coastal lines of Oman significantly vary between mountainous terrains with steep cliffs overlooking the sea as in Musandam in the northern Al Hajar Mountains or in the Dhofar Mountains, sand beaches as in the sand dunes area in north-eastern Oman, alluvial fans as in the northern side of the central Al Hajar Mountains, flat escarpments and small hills as most of the south-eastern coast of Oman. These coastal domains also have different geological history and present setting. For example, the coastline of Musandam is currently submerging below the sea at a rate of several millimetres a year due to the collision between Arabia and Eurasia, whereas most of the coastal line from Muscat to Ras Al Hadd is uplifted few millimetres every year.

1.9.1 Al Hajar Mountains

Al Hajar Mountains extend for more than 700 km from the Musandam Peninsula to the north to Ras Al Hadd to the east, forming an arc shape chain of mountain chains and long wadis that represent the backbone of the northern Oman area and make one of the largest mountain terrains in Arabia (Fig. 1.15). The chains are separated by Wadi Samail, which extends SW to NE from Samail to Muscat, hence divining the range to what is known as the Eastern Al Hajar and Western Al Hajar. The northern flanks of the Al Hajar Mountains overlook the Sea of Oman, while the southern limbs dispose water to the interior of Oman, forming long wadi channels that often end with wide sabkha or playa deposits in Central Oman. The wadis of Al Hajar Mountains, such Wadi Al Jizi and Mistal to the north and Tanuf and Muaydin to the south, historically led to the formation of many oases and settlements, whereas the coastal lines on the northern sides of these mountains have provided bays and inlets for fishermen communities to flourish.

The geology of the Western Al Hajar Mountains is characterized by thick deposits of Mesozoic carbonates, exposed in the massifs of Al Jebel Al Akhdar and Musandam (Fig. 1.14b), and oceanic rocks of Ophiolite and oceanic deposits forming the areas in between these two massifs. The carbonate deposits form elevated domes, reaching an elevation of more than 3,000 m above sea level in Jebel Shams in Al Jebel Al Akhdar, whereas the oceanic sequences may reach a height above sea level of more than 1,500 m, as in the Ophiolite Mountains south of Al Buraimi. Al Jebel Al Akhdar comprises large geological windows, such as Wadi Mistal and Wadi Al Sahtan, that may extend for several tens of kilometres and expose Precambrian rocks of mainly glacial and marine origins.

The Eastern Al Hajar Mountains comprises two main mountain chains, the Saih Hatat and Al Jebel Al Abyad. The Saih Hatat Massif is a large deeply eroded dome, 80 km × 50 km, elongated in the east–west direction. The massif exposes metamorphic rocks of Precambrian age, Palaeozoic clastic sequences and Mesozoic carbonate deposits, forming high cliffs of hundreds of metres height. During the emplacement of the Ophiolite at the end of Mesozoic, the area of Saih Hatat was subducted to deep high-pressure zones leading to significant alteration and deformation of the rocks. In contrast, Al Jebel Al Abyad is primarily formed of early Cenozoic carbonates that are less deformed and altered. The peak of Jebel Bani Jabir in Al Jebel Al Abyad, with about 2 km elevation above sea level, represents the highest peak in the Eastern Al Hajar. The Cenozoic carbonate units contain large caves and sinkholes, like Hawyat Najm, Majlis Al Jinn and Tehri Cave.

The Ophiolite exposures appear as separate blocks on the northern and southern flanks of the Western and Eastern Al Hajar. One of the largest of blocks is exposed around the Wilayat of Samail; hence, the Ophiolite of Oman is often known as the Samail Ophiolite after this large exposure. The largest exposure of the oceanic sediments is present in the Hamrat Al Duru, the Hawasina Window, Jebel Al Kawr and the Batain coast, in the eastern edge of Al Hajar Mountains.

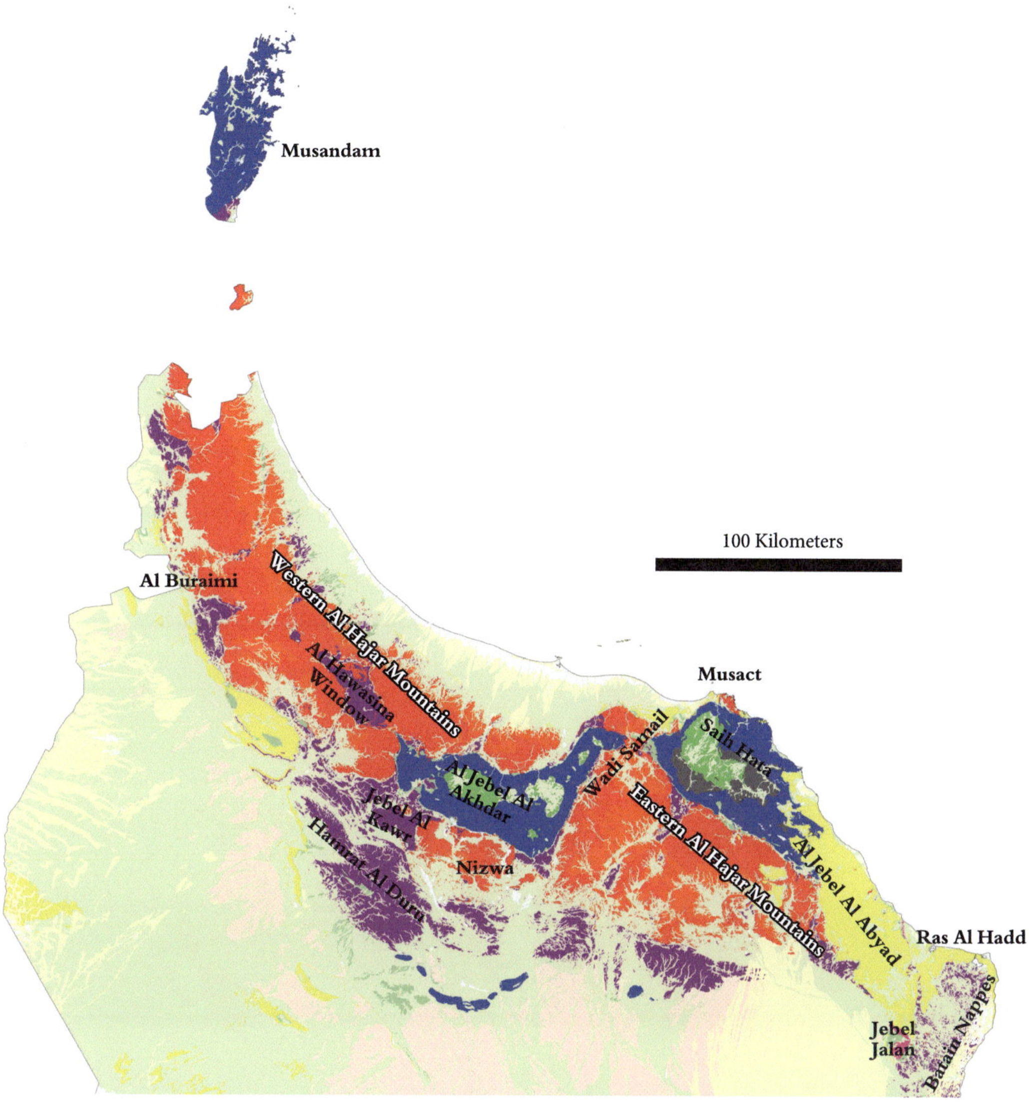

Fig. 1.15 A geological map of the Al Hajar Mountains; in summary, the warm colours (red and orange) in the north and in Masirah Island are for igneous oceanic a (Ophiolite), purple for emplaced oceanic sediments, dark pink in Jebel Jalan for Precambrian igneous basement rocks, dark green for Precambrian sediments, black for Palaeozoic deposits (in Saih Hatat), blue for Mesozoic formations and light colours (including yellow, blue and red) for Cenozoic deposits

1.9.2 Al Huqf Mountains and Masirah Island

Al Huqf area forms the eastern part of Central Oman. The area extends for about 200 km north-south and 75 km east-west. Generally, the area is covered by small escarpments and small hills above flat sabkha plains and sand dunes. This area exposes a large window of Oman's geological heritage, with Precambrian basement rocks exposed to the north, Precambrian sediments covering the eastern part, Palaeozoic sediments forming the central part and Mesozoic carbonates exposed on the eastern edge.

The Ophiolite rocks of Masirah Island and Ras Madrakah provide a unique opportunity to understand the development and composition of the crust and upper mantle of the Indian Ocean.

They are often surrounded and capped by light-colour Early Cenozoic limestone beds (Fig. 1.16).

1.9.3 Dhofar Mountains

The Dhofar Mountains are often divided into three main parts. From east to west, these are Jebel Samhan, Jebel Al Qara and Jebel Al Qamar (Fig. 1.17). Jebel Samhan and its coastal plains that extend from Hasik to Mirbat is about 75 km long. The plain exposes crystalline basement rocks of Precambrian age (also exposed in Juzor Al Hallaniyat). The cliffs of Jebel Samhan are formed of Early Cenozoic carbonates. Precambrian glacial, marine and river deposits, as well as Mesozoic carbonates, are exposed on the south-western side of Jebel Samhan. The mountain rises up to an elevation of more than 1,600 m above sea level. Jebel Al Qara and its plain, known as the Salalah Plain, form the central part of the mountain chain. The plain is predominantly covered by Late Cenozoic clastic rocks deposited mainly by old and recent alluvial fans, whereas the cliffs are formed of Cenozoic carbonates that reach about 1,000 m above sea level in the western side of the mountain. Towards this side, Palaeozoic and Mesozoic sediments are exposed. On the northern side of the mountain, the wadis flow towards the desert plains and Rub Al Khali. Jebel Al Qamar forms the south-western part of the Dhofar Mountains. This part directly overlooks the Indian Ocean

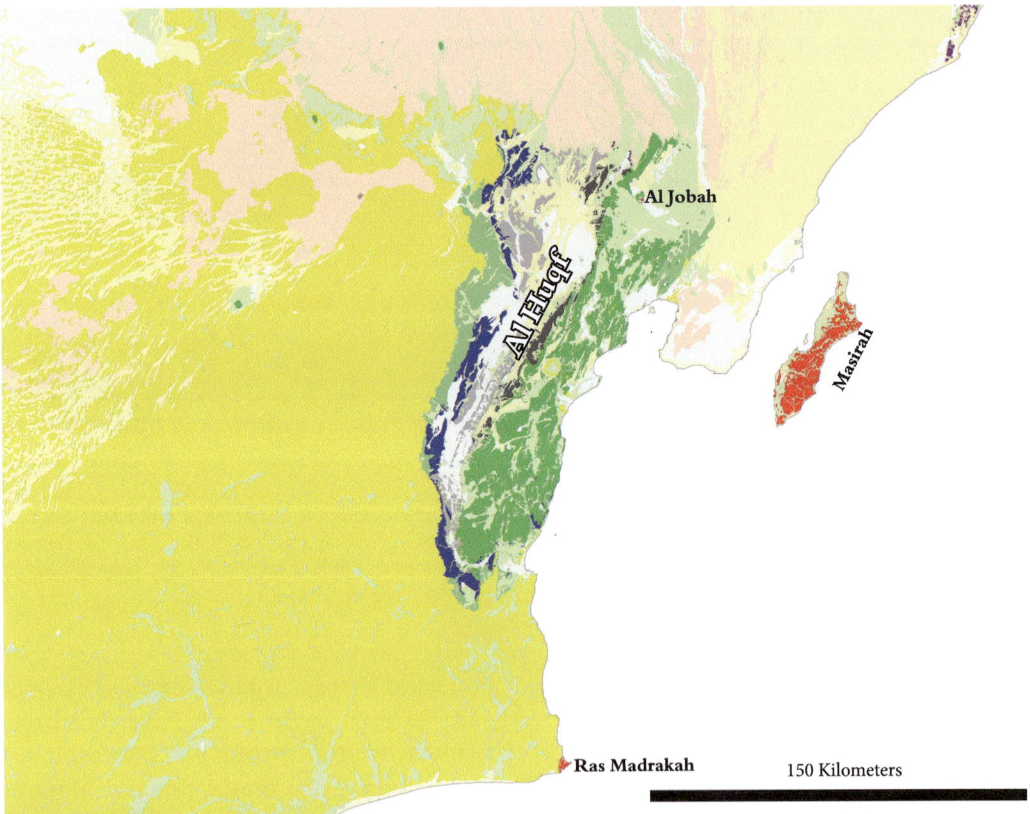

Fig. 1.16 A geological map of the Al Huqf Area; in summary, the warm colours (red and orange) in Masirah Island and Ras Madrakah are for igneous oceanic rocks (Ophiolite), pink for Precambrian basement rocks in Al Jobah, dark green for Precambrian sediments, black for Palaeozoic deposits with dark black for Early Palaeozoic and grey for Late Palaeozoic sediments, blue for Mesozoic formations, green for of Late Mesozoic carbonates in the western part of Al Huqf and light colours (including yellow, blue and red) for Cenozoic deposits

(Arabian Sea) with cliffs rising more than 700 m above sea level. The cliffs are fault surfaces that developed as a result of the Aden Rift and its associated uplift in the Late Cenozoic.

1.9.4 Desert Plains, Sand Dune and Sabkhas

The desert plain extends for an area of more than 170,000 km^2 in central and South Oman; hence, most of the area of Oman could be considered as a desert plain, as the total area of Oman is 309,501 km^2. This area does not include mountain chains and lacks major wadi channels that can cause significant erosion of the bedrock, making it a special landscape in Oman and the region. Hence, the plain remained relatively non-weathered for thousands of years, compared to the northern and southern terrains of Oman. Therefore, this area is one of the best places in the world to find meteorites that struck the earth more than 10,000 years ago and today reveals a lot of the unknown surface details of the Moon and Mars. The desert plain of central and South Oman is also characterized by bizarre rock shapes of weathered Cenozoic limestone blocks, as seen in the Rock Garden of Ad Duqm. To the north, the desert plain is bounded by a depression which extends from Um As Samim in the west to Bar Al Hikman Sabkha in the East. This depression prevents all the northern wadis to get into the desert plain.

The flat and rocky desert plain of the Central Oman Mountains has six surface-piercing salt domes that form a prominent topographic and geologic attraction in the northern and central parts of the plain (Fig. 1.17). From north to south, these domes are Qarat Al Milh, Qarat Al Kibrit, Jebel Majayiz, Qarn Alam, Qarn Nihayda and Qarn Sahmah. The domes are in rounded to elongated shapes, and they form an isolated ridge that dips towards the outer part of the dome.

Sand dune areas develop in different parts of Oman. The largest bodies of sand are found in the Ash Sharqiyah and Al Wihibah sands in north-east Oman and the Ar Rub Al Khali Sand on the south-western side of Oman. The former is rich in carbonate grains that reduce from south to north to have more quartz content in the north

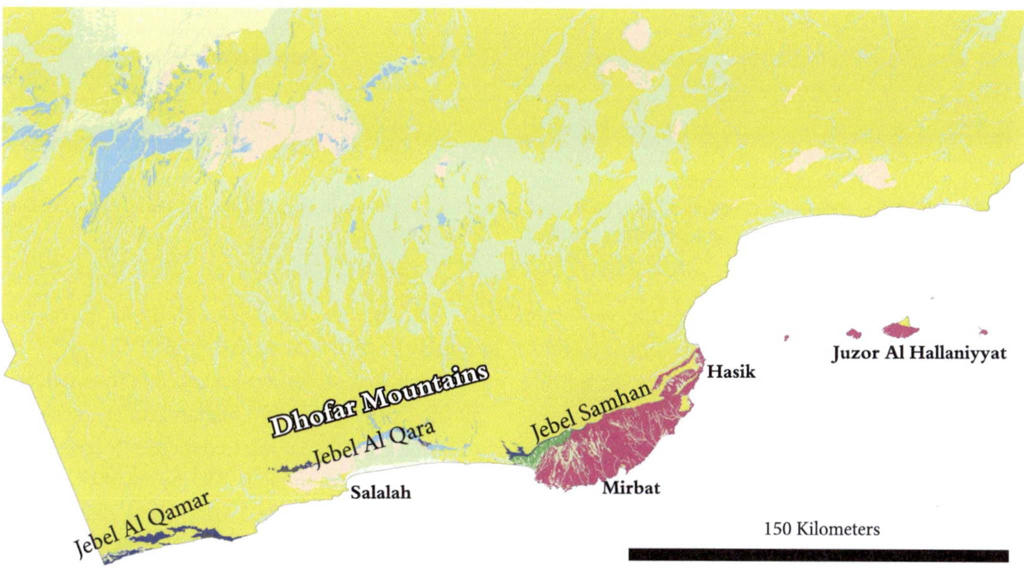

Fig. 1.17 A geological map of the Dhofar Mountains; in summary, pink represents Precambrian basement rocks, dark green for Precambrian sediments, black or grey for Palaeozoic deposits, blue for Mesozoic formations, green for of late Mesozoic carbonates in the western part of Al Huqf and light colours (including yellow, blue and red) for Cenozoic deposits. The dark yellow colour represents Early Cenozoic carbonates

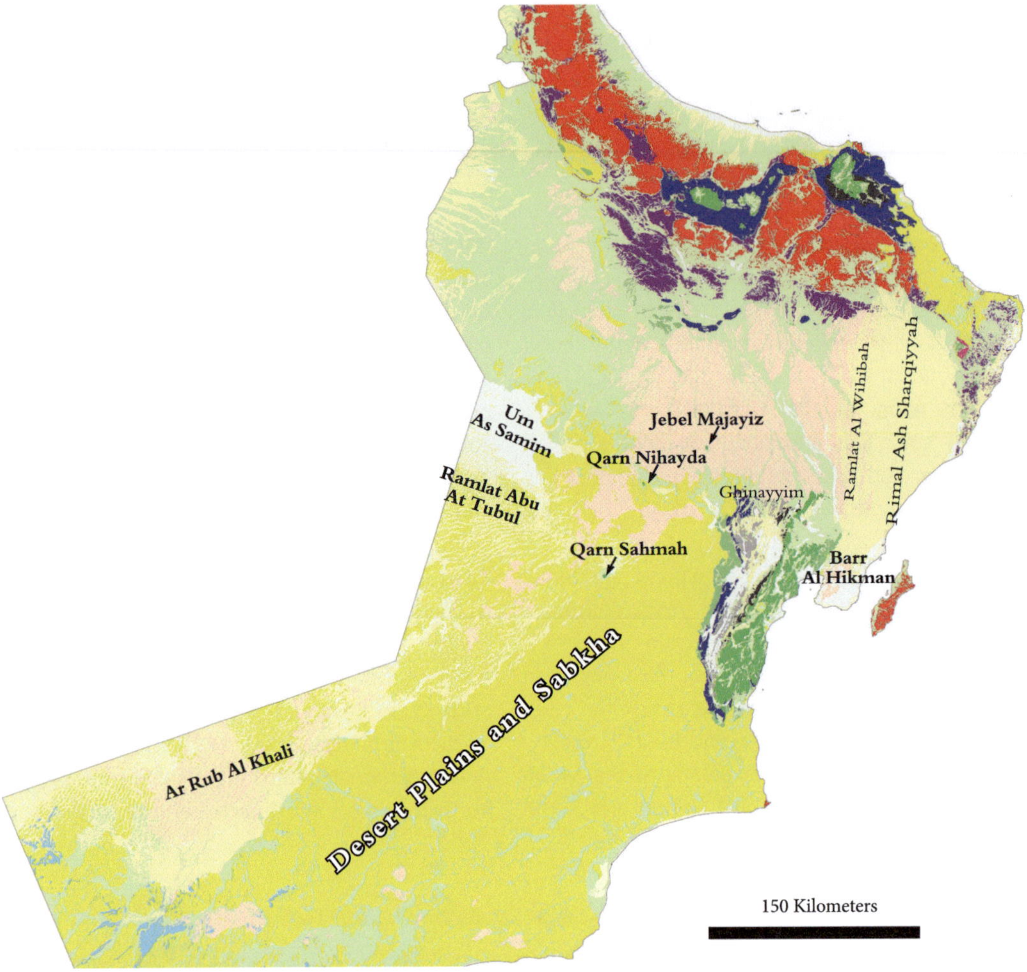

Fig. 1.18 A geological map of the desert plain in central and South Oman, dark green for Precambrian sediments exposed in the salt domes in Central Oman. The blue colour in the south is for lacustrine deposits of limestone and conglomerate, the light yellow is for sand dunes (aeolian deposits), the light red is for reddish conglomerate and siltstone, light blue, particularly exposed in Um As Samim, and light green is for alluvial gravel and sand. These deposits are Late Cenozoic in age, whereas the dark yellow is mainly for carbonates from Early Cenozoic

and was mainly deposited during the previous glacial periods in the last few 100 s of thousand years when strong winds were blown sand from south to north to form mega linear sand dunes. The latter, Ar Rub Al Khali, is one of the largest sand dunes in the world and covers an area of about 300 km by 80 km in southwest Oman. In Oman, most of the Ar Rub Al Akhali Sand is formed of barchan or star dunes, or combinations of sand geometries that are difficult to categorize. The sand dunes are often separated by sabkhas, with abundant gypsum and gravel plains. Between the sand dunes, old deposits of lake sediments are found occasionally.

A long line of dunes also extends for more than 250 km from the Empty Quarter, known at this area as Ramlat Abu At Tubul, to the eastern coast of the Central Oman Mountains. The Wilayat of Mahout is also characterized by white sand dunes, often known as the Sugar Sand. The sand dunes have high contents of carbonate grains blown out from the south-eastern side by the southwest monsoon. The northern part of the sand dunes in this part of Oman is called Ramlat Ghunayyim. (Figure 1.18)

2 The Precambrian Rocks and Life in Oman

Domes known as stromatolites were formed by bacteria: one of the most prominent features of life during the Precambrian. These domes are from the Al Huqf area, east Oman.

M. H. Al Kindi, *Evolution of Land and Life in Oman: An 800 Million Year Story*,
https://doi.org/10.1007/978-3-319-60152-6_2

2.1 Earth During the Precambrian

The Cryptozoic Supereon, or what is informally and widely known as the Precambrian, represents the first and largest time span of earth's history. It is divided into three eons (Fig. 2.1): these are the Hadean, the Archean and the Proterozoic. It extends from the formation of earth some 4,600 million years ago to 541 million years ago, and therefore spans 88% of the earth's age. However, this supereon is relatively unknown due to the poor record of rocks and fossils from this time. Its rocks have been heavily weathered, eroded, metamorphosed or buried and recycled over time. Our understanding of the environmental changes and evolution of life during this time is, therefore, limited.

During the Hadean, the earth was as hot as the underground hell of its interior. It takes its name from Hades, the God of the underworld from Greek mythology. It had just become a terrestrial planet separate from the rest of the universe after being part of a nebular cloud following a supernova, a name given to the explosion of a star and the emission of huge radiations. The nebula contracted and began to rotate and accelerate. The contraction led to the fusion of hydrogen nuclei into helium in the centre of the nebula which generally lacks angular momentum. Planets, like the earth, formed in other parts of the nebula, particles collapsed into separated rings that later formed rotating planets as these particles clumped together. The earth possibly originally grew bigger by gravitational accretion of smaller bodies around a dense core.

In fact, many scientists believe that the earth was at least 2% larger than its current size, until a planet the size of Mars collided with it more than

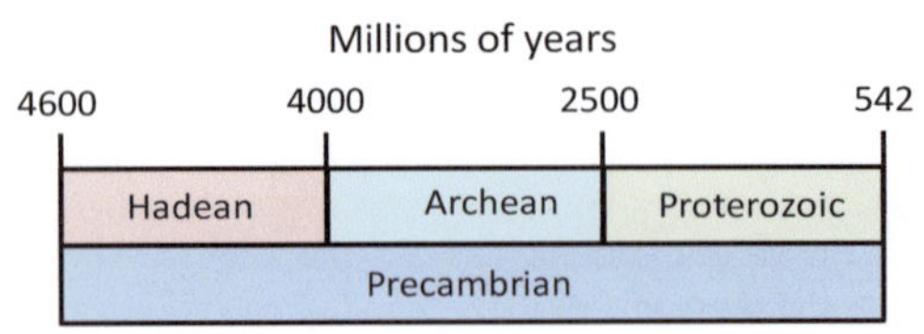

Fig. 2.1 Three eons of the Cryptozoic Supereon or the Precambrian

4,000 million years ago and removed a chunk of the earth to produce the moon. Some claim that this collision was probably responsible as well for the tilting of the earth's axis. Most of the earth was molten, and terrestrial bodies were frequently smashing its surface. It was not possible for life to start under such harsh conditions, but the foundations for life were being prepared. Water was filling its surface as the cold and solid crust was developing; while the hot, heavy elements were coalescing in its interior.

In the Archean (from the Greek arkhē meaning 'beginning'), the outside layer of the earth had already cooled and formed a crust which hosted the first organisms on the planet. The atmosphere was probably still hot, as greenhouse gases were continually warming the planet. The internal heat flow was much larger than today. However, small crustal plates were formed and probably scattered on the surface of the earth (Fig. 2.2). Although heat convection cycles were possibly more than double the heat flowing today, life managed to emerge at least 3,500 million years ago. It was prokaryotic bacteria, simple, single-celled organisms. An example is cyanobacteria, which are probably the only known prokaryotes that produce oxygen from photosynthesis. Their fossils are found in places like South Africa, Australia and Canada. They released significant amounts of oxygen and consumed large quantities of carbon dioxide (CO_2). Some 3,200 million years ago, the amount of oxygen in the atmosphere spiked, while the level of greenhouse gases fell—significantly reducing the temperature of the earth. This was possibly supported by the smaller and less luminous sun—smaller by about 20% than its current size and around 80% as luminous as today.

About 3,000 million years ago, the earth had already cooled. However, this did not last for long as the oxygen levels began to drop 2,800 million years. We cannot be sure of the reason, but such cases are often related to increased ultraviolet light and solar flares, or the consumption of oxygen during the decomposition of bacteria. The earth started to become hot again and the number of oxygen-producing photosynthesizers reduced,

Fig. 2.2 The evolution of the earth throughout the three eons of the Precambrian

opening doors to another type of microorganism known as methanogens to flourish. Unlike cyanobacteria, methanogens produce methane as a by-product. They feast upon dead bacteria and pump significant amount of methane into the atmosphere. Methane has over twenty times as great a warming effect on climate compared to CO2. Between 2,800 and 2,500 million years ago, the earth was warm again.

About 2,500 million years ago, the Proterozoic began. The term comes from the Greek word proton meaning first. At the start, the earth underwent another transformation phase. The percentage of oxygen in the atmosphere started to increase again as another type of prokaryote known as methanotrophs proliferated. Methanotrophs metabolize methane, which was abundant in the atmosphere. As methanotrophs became widespread, the percentage of methane significantly reduced, allowing more sunlight to reach the earth and also allowing the renewed spread of oxygenic photosynthesizers. This resulted in an increased oxygen percentage in the atmosphere, which had by then reached 2% of its present value. This increase was probably supported by a reduction in the amount of iron and sulphide that oxidize by trapping oxygen from the atmosphere, and by the lower luminance of the sun. By 2,200 million years ago, the earth started to enter its first glaciation period, and a great portion of the earth's surface was covered by ice.

The vitality of our planet and its systems keeps it in continuous transformation (Fig. 2.2). When the earth entered its first glaciation period 2,200 million years ago, an estimated 99% of its organic matter was destroyed. As it died, thick layers of dead microbes were deposited. While the microbes decomposed, large quantities of oxygen were consumed. Consequently, the methanogens that live on dead organic matter became widespread again and pumped methane and CO2 to the atmosphere. The size of the sun and volcanic activity would have both increased as well. About

250 million years later, the snowball earth started to melt, and the earth began warming again.

Primitive microbes have an astonishing capacity to change the climate of the earth. They are present in massive quantities on land and sea and have played a major role in refining the earth's climate during the early periods of life. They consume a variety of substances including rocks, minerals, enzymes and gases, and convert them to different elements which support the continuation and evolution of life on earth. When the ice started to melt and the oxygen level and temperature increased 2,100 million years ago and beyond, new types of organisms evolved, known as eukaryotes. In contrast to a prokaryote, a eukaryote cell contains a nucleus. The evolution of eukaryotes is considered a major evolutionary step in earth's history. The oldest known fossils of eukaryotes are found in different places in Africa, North America and Australia. Eukaryotes probably resulted from the merging of two primitive cells, one of which formed the mitochondrion—the power plant or centre of energy production in the cell. Mitochondrion is found today in all eukaryotic cells that make up animals, plants and fungi. Its emergence had a dramatic impact on the evolution of complex life.

The earth underwent four major periods of glaciation during the Proterozoic; three of them at the end of the eon, from 1,000 to 541 million years ago. The continental plates had then coalesced in a supercontinent called Rodinia that was centred around the equator and was surrounded by a vast superocean known as Mirovia. During the formation of Rodinia, crustal plates converged leading to the building of major mountain ranges along convergence zones. This tectonic process was accompanied by volcanic activities that led to the formation of igneous intrusions which expanded the sizes of continents.

When Rodinia started to break apart as a result of plate tectonics—about 750 million years ago —vast tracts of land were affected by erosion and weathering, exposing the rocks to the oceans. Since rock is rich in silica, great amounts of silicates were released, both into the atmosphere and into the oceans. They reacted with carbonates in the sea water, changing the chemical composition in a reaction capable of lowering temperatures on a global scale. It was this factor along with the production of oxygen through microbial activity and the increase of volcanic ash obscuring the sunlight that led to the three glaciation cycles at the end of the Proterozoic (Neoproterozoic). The other part of the cycle—global warming and the melting of ice caps—was a result of the increase in methane and CO2-releasing organic matter, like methanogens, and the pumping of CO2 into the atmosphere through volcanic activity.

These cycles of glaciation and global warming during the Proterozoic acted like a screening process for the genes of single-celled organisms. These organisms were constantly evolving during these harsh conditions. Many became extinct, and many others adapted and grew. It was like a testing ground, preparing earth for complicated life forms to bloom. Almost four billion years of experimentation, trial and error were required to create the right atmospheric and terrestrial conditions to sustain the emergence and stability of complex forms of life. Throughout this long period, the earth went through extreme conditions and cycles of temperature. These were eventually balanced by the wonderful bioengineering of microbes which produce or consume oxygen and greenhouse gases. During this period, the proportion of oxygen increased in both oceans and the atmosphere, the ozone layer which absorbs most of the sun's harmful ultraviolet radiation was formed, and sufficient amounts of silica and calcium needed to build shells and skeletons were produced by cyanobacteria. These factors resulted in the formation of the first multicelled life forms 635 million years ago, and the oldest evidence of their presence is found in Oman.

2.2 Oman During the Precambrian

The Arabian Plate started to form into its final shape when microcontinental plates coalesced nearly 850 million years ago (Fig. 2.3). The eastern part of Oman was separate from the western side. The suture zone of the two parts runs along the central and eastern parts of Oman. The suturing process resulted in the formation of many intrusive igneous dykes—narrow bands of rock that push between existing layers. They are clearly visible at Jabal Qahwan in the Wilayat of Jaalan Bani Bu Hasan, in the area of Al Jubah in Mahut and along the Mirbat Plain in the Dhofar Governorate (Figs. 2.4, 2.5 and 2.6). These dykes and the surrounding granitic and metamorphic layers are more than 800 million years old and therefore represent the oldest rock formations in Oman.

During the Precambrian, Oman and the rest of Arabia were positioned close to the equator (see Fig. 1.5). Despite the fact that this area is usually one of the hottest places on the surface of the earth, Oman—and probably the whole globe—was actually covered by ice sheets during at least three successive periods at the end of Precambrian. The rocks deposited during these glaciations periods can be seen on the Mirbat Plain and the wadis (valleys) of Al Jabal Al Akhdar, such as Wadi Bani Kharus and Wadi Mistal (Figs. 2.7 and 2.8). They are igneous and metamorphic pebbles, such as granite and schist, that date back more than 2,500 million years and they are, therefore, considered among the oldest pieces of rock found in Oman. Although it is likely they were formed in neighbouring subcontinents, these pebbles were carried by ice and redeposited in Oman about 700 million years ago. They represent continuous successions of glacial-origin rocks (rocks deposited in glacial conditions) and sandstone (rocks deposited by rivers or under ice sheets), and in some cases, limestone beds that were deposited in marine conditions during the interglacial periods (warmer global periods between glacial cycles).

Ice sheets could transport rock boulders for thousands of kilometres away from their source.

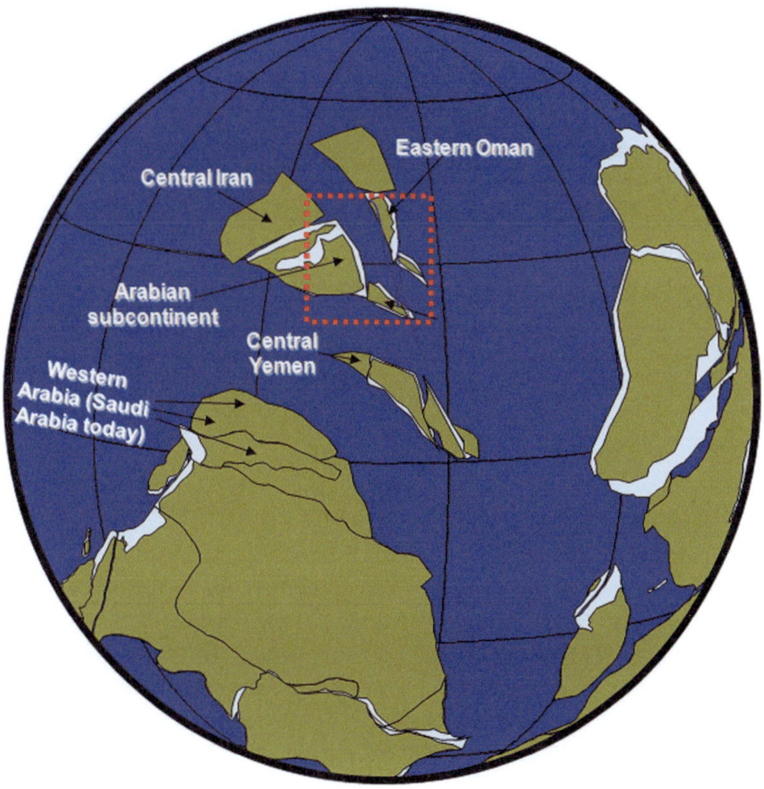

Fig. 2.3 Formation of the Arabian Plate about 850 million years ago. Before accretion, the plate was divided into two major parts, the eastern (today the eastern side of Oman) and western segments

Fig. 2.4 Igneous and metamorphic layers that represent the basement of the Arabian Plate. The intruded igneous dikes (dark in colour) cut the basement rocks in Jabal Qahwan in the Wilayat of Jaalan Bani Bu Hasan. Some of these dikes date back to more than 800 million years, therefore the rocks they cut (in this picture these appear green to grey in colour) must be much older in age than 800 million years

Fig. 2.5 Granodiorite (a type of granite) basement rocks that form the foundation of the Arabian Plate can be seen in Al Juba in the Wilayat of Mahut. They were formed during the accretion of Arabia in a magmatic arc and small continental terrains along the south-eastern margin of Oman more than 750 million years ago. The process of accretion was probably completed 720 million years ago

They often carried these rocks across different continents, and when the ice melted, the transported rocks were either deposited on land or in the sea. Some of the glacial boulders in Mirbat and Al Jabal Al Akhdar were recently dated to around 2,550 million years old and belong, therefore, to the Archean Eon. Their origin could be the Indian Subcontinent, which was formed at this time—about 3,000 million years ago—or perhaps they were transported from Saudi Arabia, Yemen and neighbouring areas (Fig. 2.9). The rocks were then carried by ice sheets more

Fig. 2.6 Igneous basement rocks from Hadbin in the Wilayat of Sadah

Fig. 2.7 On the left, scratches and grooves made by glaciers as they moved on the underlying rocks beds, most likely from south to north, during Precambrian times in Mirbat. North is at the top. On the right, a dropstone of about 10 cm length—an isolated fragment of igneous rock dropped by glaciers on the underlying fine-grained bent strata

Fig. 2.8 Mixture of clasts (mainly igneous and metamorphic) deposited by Precambrian glaciers in Mirbat, more than 700 million years ago. The orange dykes (known as Neptunian Dykes) which extend for long distances and cut across boulders are formed by the overpressure of the overlying ice sheets

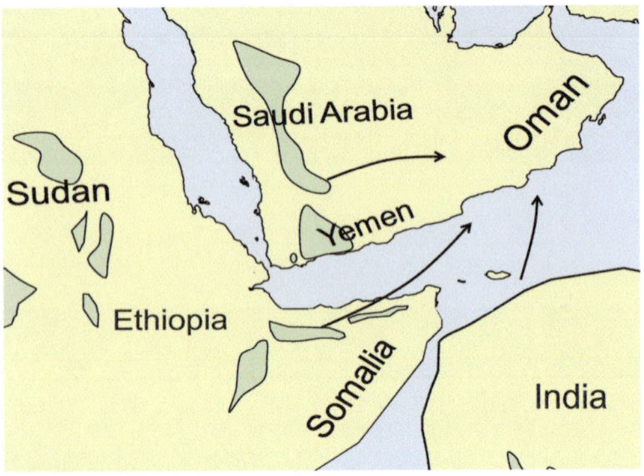

Fig. 2.9 Map roughly represents the location of the Indian Subcontinent at the end of the Precambrian, redrawn and modified from Rieu et al. (2007). It shows the locations, in green, of rocks older than 1,000 million years which might be a source of the igneous and metamorphic clasts that were carried and redeposited by ice sheets around 700 million years ago when Oman was covered by ice. The oldest rocks are found in India, Yemen and Saudi Arabia. Their age is estimated to be between 2,500 and 3,000 million years old

than 700 million years ago. They do not represent an independent geological formation in Oman, but are depositional ingredients of the glacial formations. With the exception of fragments of meteoric rocks, found on the desert plains, these are considered the oldest rocks in Oman.

The glacial periods of the Precambrian ended in Oman about 635 million years ago when global temperatures and volcanic activity increased. Worldwide sea level rose, depositing thick layers of carbonate rocks. Within them, different forms of simple life lived. As they died, organic-rich layers of carbonate rocks were deposited. Today, these layers are the most important source of hydrocarbons in the oil and gas fields of Oman. They form thick and prominent outcrops in the Al Hajar Mountains in North Oman and in the Al Huqf Province of Central Oman (Figs. 2.10, 2.11 and 2.12).

After the deposition of organic-rich carbonate layers, about 541 million years ago at the Precambrian–Palaeozoic boundary, the sea receded and great quantities of salt precipitated, forming isolated salt basins in north and South Oman. On the surface, these deposits can be seen as salt domes in Central Oman, such as at Qarat Al Kibrit, Qarn Alam and Qarn Sahmah in the Wilayats of Haima and Adam. The domes formed as the salt gashed from a depth of around 4–5 km and pushed the overlying rock layers to surface, forming circular salt domes on the flat plain of the desert. These domes contain a mixture of many rocks that are mainly deposited during the Precambrian and Palaeozoic times. The salt has been mined since ancient times. Pictures of the domes taken in the 1950s show tribal workers who skilfully formed regular blocks of the mined salt which were transported

Fig. 2.10 Carbonate deposits in the western part of Al Hajar Mountains. They formed in a shallow marine environment more than 540 million years ago and are the most important rocks in the petroleum system of Oman, because they form the source rocks for oil and gas for more than 80% of the hydrocarbons found in the subsurface layers of Oman

Fig. 2.11 Buah Dome in Al Huqf area in the eastern part of Central Oman. These deposits are more than 540 million years old and were laid down in shallow seas at the end of the Precambrian. The dome contains Precambrian rocks in its centre and Palaeozoic strata around the rim. It is about 6 km from north to south and is cut through the middle by a major fault oriented north–south

Fig. 2.12 Precambrian and Palaeozoic carbonate rocks in Al Jabal Al Akhdar (Wadi Bani Awf) are separated by a major time gap (where the dotted black line is) of erosion or non-deposition, equivalent to about 290 million years, from about 550 million years old rocks to 260 million years old. The Precambrian rocks have been significantly deformed before the deposition of the Late Palaeozoic rocks

by camel to their villages (Figs. 2.13 and 2.14). As part of the subsurface, salt layers have played a major role in the formation of hydrocarbon traps in the petroleum fields of Oman.

2.3 Organisms Living in Oman During the Precambrian

2.3.1 Bacteria

Bacteria and other types of prokaryote appeared on earth more than 3,500 million years ago. They represent the first manifestation of life on earth. The oldest evidence of their presence in Oman dates back to more than 700 million years ago during the Proterozoic. This evidence can be seen in the Precambrian carbonate beds of the Mirbat Plain (Fig. 2.15). These microbial beds of bacteria are most likely formed by a type of bacteria known as cyanobacteria which forms large colonies.

Cyanobacteria form layered microbial mats that commonly grow bigger as both isolated or connected accretionary bacterial domes known as stromatolites (Fig. 2.16). These domes were very common in Precambrian times. Through photosynthesis, they produced large amounts of oxygen which played a vital role in converting the ancient atmosphere into an oxidized one, contributing to the emergence of more complex organisms on our planet (Fig. 2.17).

In comparison with Precambrian times, stromatolites today have significantly declined in abundance, size and diversity. The size and shape of stromatolitic domes are directly linked to the environment in which they form. Large domes often form in open marine environments where waves and tides are active, whereas smaller and flatter domes form in restricted marine environments and coastal areas. It is important to note that Proterozoic stromatolite domes could also be formed by green algae.

Fossils stromatolites are widespread among the rocks in Oman deposited in the Proterozoic Eon between 640 and 540 million years ago (Fig. 2.18). They are situated in the outcrops of the Al Hajar Mountains and Al Huqf area. They can also be seen in the salt domes in Al Dhakhiliah and Al Wusta, as in Qarn Sahmah and Jabal Majayiz. The absence or lack of other living organisms as predators or that could

Fig. 2.13 This photograph is taken in the 1950s by George Laurnace. It shows a tribal worker mining ancient salt deposits to use them mainly as food for them and their animals. Some of these mines are dated to more than 6,000 years from present

compete for food supported their spread and abundance. Stromatolite rocks frequently also contain fossilized wave ripples that point to the ancient shallow seas where the stromatolites once lived.

Stromatolites appear in different shapes (Figs. 2.19 and 2.20). They could be domal, columnar or branched. The organic matter produced by the deposits of stromatolite layers is considered one of the most ancient and valuable sources of hydrocarbons in Oman. Moreover, the fossils of stromatolites are useful in understanding the environmental changes on the planet in the Precambrian time. They are also used in determining and comparing the age of rocks. The number of fossilized stromatolite domes fell sharply after the Precambrian. Fossil domes can be found in smaller numbers in the Palaeozoic, Mesozoic and Cenozoic rocks of Oman and in some parts of the world, such as Australia and the Bahamas, they still live along some beaches today.

Fig. 2.14 There are six surface-piercing salt domes in the desert plain of Central Oman. Ancient salt mines within the domes have been in continuous use until recent times. This photograph was taken in Jabal Sahmah in the Governorate of Al Wusta in the 1950s by George Laurance

Fig. 2.15 This bed of alternating dark and light colours in Mirbat (indicated by the green arrow) might be the oldest known life form that ever existed in Oman, dated to more than 700 million years old. It represents a microbial mat that was deposited in the interglacial periods of the Proterozoic Eon, when global temperatures rose between the glacial periods. This organic bed is underlain by granitic rocks that represent the basement of the Arabian Plate (red arrow)

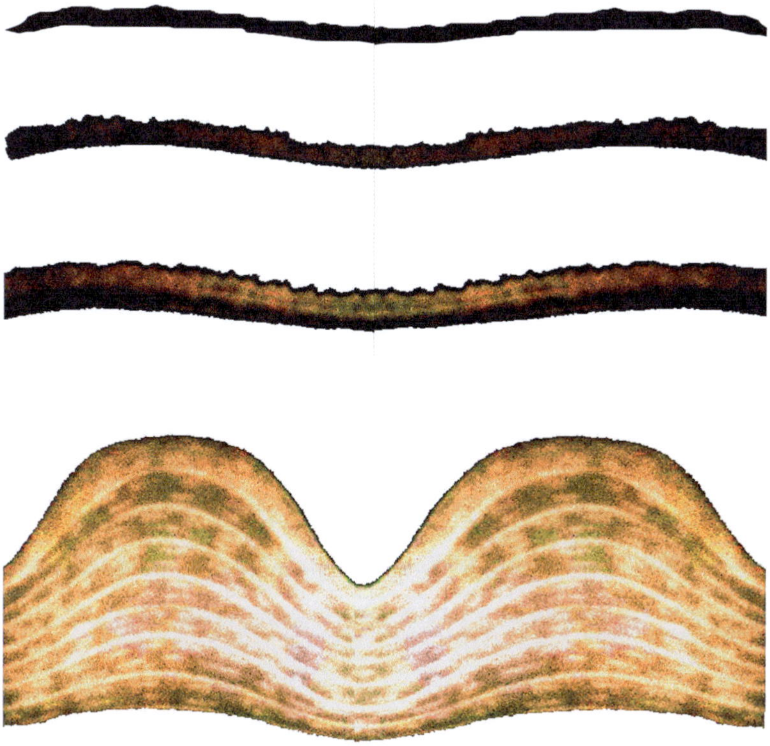

Fig. 2.16 Illustration of the stages of vertical stromatolite growth from top to base. They grow in shallow seas, first as microbial filaments which trap and bind sediments from the sea. Later, another bacterial mat appears on top of the sediment. The layers can then grow rapidly as domes and might form connected domal structures with time. These connected domes grow later as one large dome that might reach many metres in height and width. The shape of the internal structures of a stromatolite dome, stromatolite laminas, records the evolution stages of the dome and can display many details about the biological processes and microbial growth of the stromatolite dome

Today, different types of bacteria live in all parts of Oman, particularly in lakes—often under harsh conditions of high salinity, sour or brackish water and high temperatures. For example, algae, bacterial mats and columns cover the bottoms of some hot springs (Fig. 2.21) that gush along major cracks or faults extending from the western side of the Wilayat of Ar Rustaq (Al Kasfah Spring, where water temperature is about 50°C) to the eastern side of the Wilayat of Bawshar (Al Hammam Spring, where water temperature is about 60°C). Bacteria and algae live in the white carbonate lakes that form in the Ophiolite mountains as well, such as Wadi Al Jizi (Fig. 2.22). Bacteria also flourish in the pink lakes, which are present in the Wilayats of Al Jazir and Muhut (Fig. 2.23). These lakes formed in salt playa or *sabkha* and along shoreline tongues or lagoons that were isolated from the sea by coastal sand dunes. They have an extremely high salt content, particularly during summer because of increased water evaporation rates. The high salinity and temperatures in the pink or red lakes promote the existence of astronomical numbers of microscopic salt-loving bacteria, known as Halobacteria or Archaebacteria (Fig. 2.24). They produce a red pigment from which the lakes take their colour. Halobacteria can flourish in very harsh environmental conditions of up to 30% salinity and extremely warm water. The pink lakes also contain green algal mats that feed on salt, too.

Fig. 2.17 On top: sequence of life forms in Oman throughout the four periods of life since the Precambrian. The stromatolites appear in the Precambrian and continue to live up to date, therefore they are one of the longest living forms of life on earth. They were more abundant and diverse during the Precambrian. They converted earth's atmosphere in the Precambrian to an oxygen rich. Below: stromatolite domes proliferated in the shallow seas of the Precambrian, as illustrated in the Story of Life in Oman film. These domes were the most important form of life at the time. They photosynthesized sunlight to produce oxygen

Fig. 2.18 Continuous layer of stromatolite domes (black arrow) in the Al Huqf area with wave ripples in the rocks below them. The stromatolites develop on the top layer and form domal structures that start small and grow bigger. If the appropriate growth environment is maintained for a long period, stromatolites will grow into massive structures

2.3.2 Acritarchs

Among the organic fossils found in the Precambrian and Palaeozoic rocks of Oman are the acritarchs. In principal, this classification of organisms includes any small organic-walled microstructure of unknown biological classification (Fig. 2.25). The name is a combination of the Greek akritos and arch, meaning 'confused origin', therefore indicating the unknown class of these organisms. They could represent single cells of prokaryotes and eukaryotes, or planktonic organisms that floated in the sea before being deposited on the sea floor after their death. It is common that organisms with unknown biological source are initially classified as acritarchs, but once they are properly identified, they get reclassified according to their real origin. Acritarch fossils possibly appeared on earth around 3,000 million years ago and are therefore considered among the oldest microfossils that ever lived. They are useful in determining the precise age of Precambrian and Palaeozoic rocks in Oman, being considered one of the most widely spread organisms during these periods. They are often used to correlate the subsurface rocks of the oil and gas fields in Oman.

2.3.3 Ancient Marine Sponges

Sponges are part of the phylum Porifera deriving their name from the Latin porus meaning 'pore' and –fer meaning 'bearing'. Although part of the kingdom Animalia, they have a very simple structure with bodies full of tubes and holes. They could be considered as a bundle of multicells that stick together. They do not have any digestive, nervous or circulatory systems, and they are unable to move. They depend on the flow of water that percolates through their bodies, carrying food and oxygen and removing their

Fig. 2.19 Stromatolites appear in different shapes and sizes in the outcrops and subsurface rocks of Oman. These columnar stromatolites were found while Petroleum Development Oman was drilling an oil well. Each sample is about 15 cm long. The internal dark and bright layers represent alternation between dark organic-rich layers and bright sediment-rich layers. The sediment grains were trapped by the dark microbial filament. The organic-rich layers of stromatolites are a major source of hydrocarbon in Oman

Fig. 2.20 Large stromatolite dome, about 4 m long, in the Precambrian rocks of Oman, in the central part of Al Huqf area. This dome is about 560 million years old

Fig. 2.21 Small green columns and the orange mat that grow at the base of the Al Kasfah hot spring in Ar Rustaq are bacterial and algal communities that live in environments reaching 55 °C. The photograph shows an area of around 3 m in width

Fig. 2.22 White carbonate lakes formed as olivine-rich Ophiolite rocks react with freshwater and CO_2. This process reduces the amount of CO_2 in the atmosphere and forms many cracks and veins of magnesite in the Ophiolite rocks. Bacteria and algae form microbial mats in some of the pools that spread across the northern and southern sides of the Al Hajar Mountains

waste. However, this very simple animal is the very first multicell organism that appeared on earth.

In February 2009, global news channels broadcast an important scientific discovery when a team from the University of California discovered the oldest evidence of sponges in the world. It was found in rocks from South Oman. It represents a chemical biomarker known as 24-isopropyl cholesterol. These biomarkers belong to demosponges that lived in Oman 635 million years ago. Thus, the rocks of Oman comprise, so far, the oldest known evidence of the emergence of multicell organisms. Interestingly, the philosopher Al Biruni, around 1,000 years ago, was probably the first to classify sponges and corals as animals because they respond when touched. The appearance of sponges on earth was the start of the emergence of many different types of animal. They evolved after billions of years of a world occupied by only single-cellular microbes, such as the stromatolite-forming bacteria. Dividing cells used

Fig. 2.23 Different bacteria-supporting lakes in Oman. The pink or red lakes and playas in the Al Wusta Governorate in Central Oman

to separate and drift apart, but around 635 million years ago, everything changed.

At the end of the Proterozoic, with high percentages of oxygen in the atmosphere as a catalyst, the divided cells started to stick together (Fig. 2.26). They formed into clumps using proteins like glue and became the first multicell organisms, like sponges. Sponges probably first

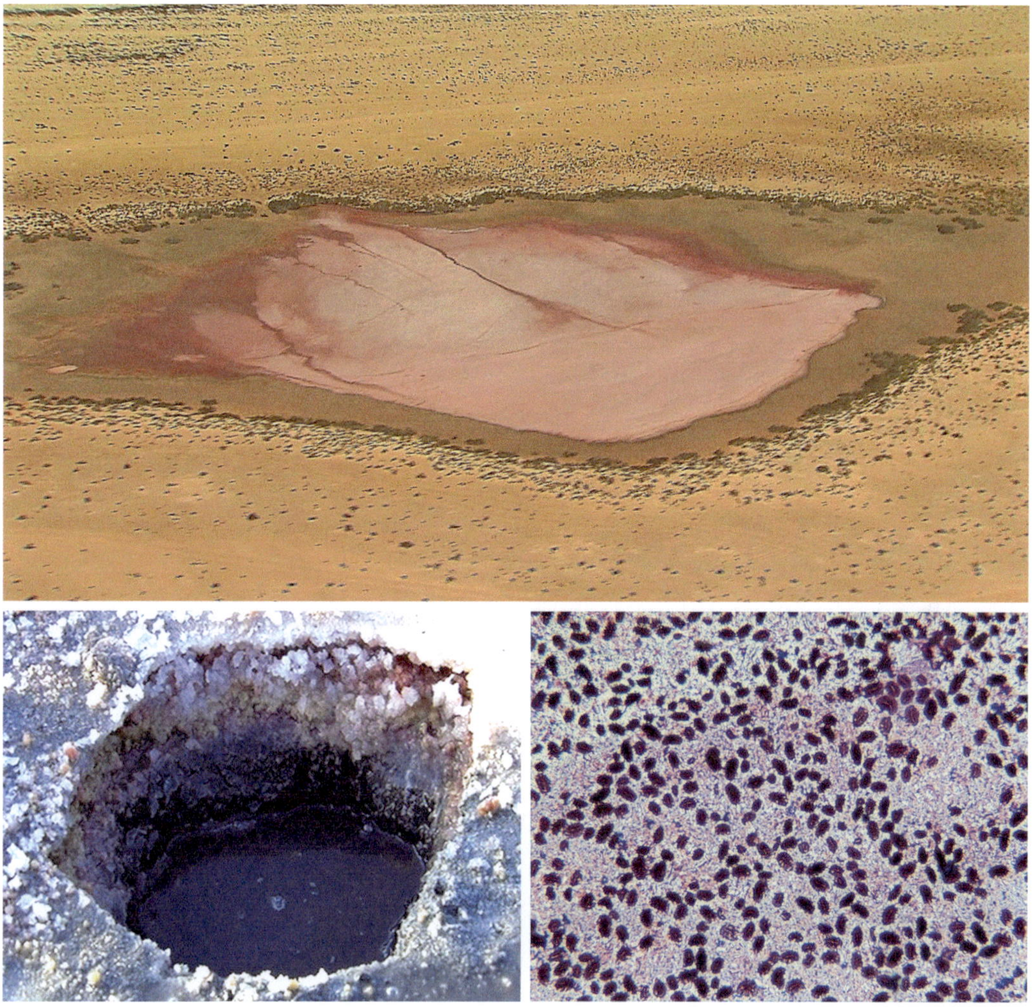

Fig. 2.24 Left: a small hole, about 30 cm wide, in the pink lake of Bar Al Hikman (shown on the top photograph), presenting a cross section of the upper layers of the lake. It shows a black layer of dead organisms overlain by a green layer of algae and fungi, followed by a pink layer of red or pink salt crystals. Cubic crystals of sodium chloride (NaCl), or food salt, cover the lake with a crust full of microscopic organisms that produce a red pigment. Right: a microscopic photograph of Halobacteria from the pink lakes in Al Wusta Governorate. This tiny bacteria is present in vast numbers and produces a red pigment that colours the water in the lakes of Al Wusta Governorate a red or pink colour

lived in shallow seas because the deep sea was unlikely to have contained sufficient oxygen. Although the climate was still relatively cold within an interglacial period, the absence of predators allowed the sponges to spread and thrive.

The discovery of sponge biomarkers in the rocks of South Oman had a global impact. It was previously thought that the first animals appeared on earth only 580 million years ago, but the revelations of 2009 unveiled a much more ancient date for the emergence of animals. Furthermore, palaeontologists across the world no longer had to rely just on visible fossil evidence to deduce the presence of organisms, but could add chemical biomarker evidence. This continues to contribute to discoveries throughout the world.

Fig. 2.25 Microscopic images of different types of acritarchs. The top row is from the Precambrian Era (Butterfield and Grotzinger, 2012), whereas the second and third rows are from the Palaeozoic Era, analysed in the laboratories of Petroleum Development Oman. Each is about 300-400 μm across

The ancient sponges living in Oman 635 million years ago most likely appeared before the last glaciation (known as the Gaskiers Glaciation) in the Proterozoic. Generally, such cold periods have not supported the evolution of complex life forms. After they ended, life flourished on earth and animals appeared. About 100 million years after the emergence of sponges at the end of the Precambrian, life had already exploded to form diverse types of complicated organisms.

Fossils of sponges can be seen in many locations in Oman. One of the most attractive sites is located in Al Jabal Al Akhdar, where many fossils of sponges from the Palaeozoic occur (Fig. 2.27). These fossils are present in the Mesozoic and Cenozoic rocks, too, particularly in the Dhofar Mountains. Fossils of sponges have also been recovered while drilling cores in petroleum wells. Today, sponges are widely spread throughout the seas around Oman (Fig. 2.28). They are no longer limited to shallow waters as they were 635 million years ago. As the level of oxygen in deep water has increased, they have become even more diversified.

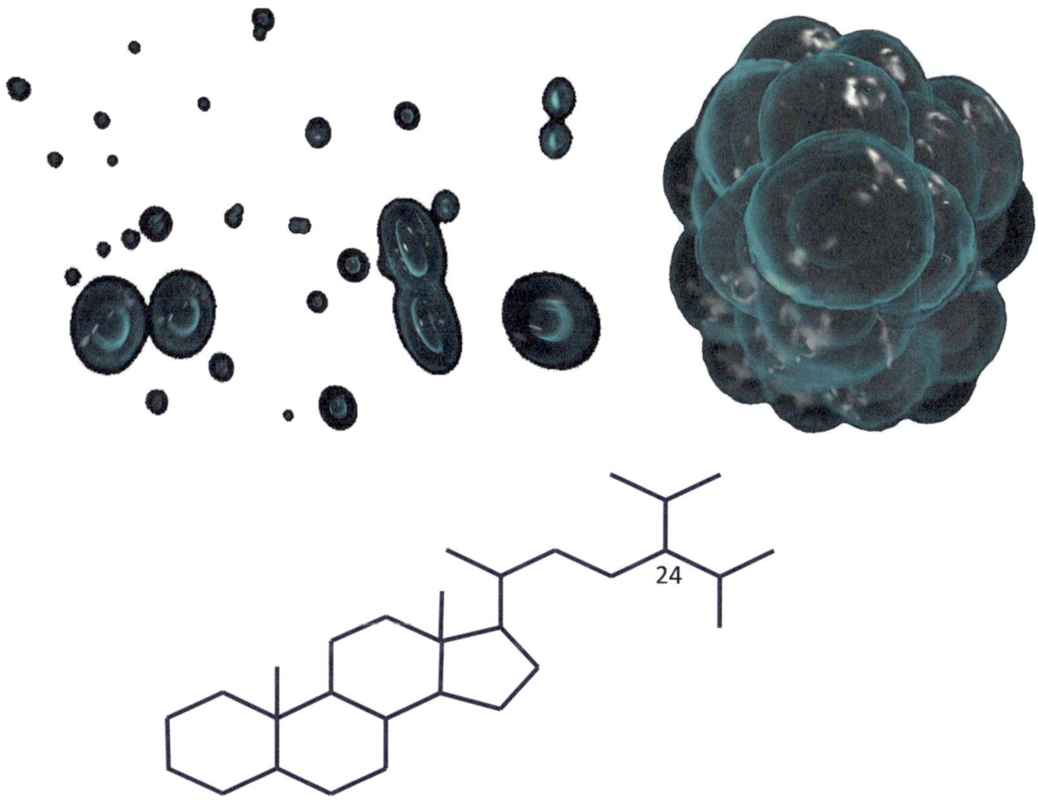

Fig. 2.26 Sponge development. On the top left: the dividing cells separate from each other until, on the top right: they started to stick together after their division, using a common and complex protein, known as collagen. As shown on the bottom right: the chemical fingerprints of sponges were found in Oman represent a form of cholesterol, known as 24-isopropyl cholesterol

2.3.4 The Precambrian and Phanerozoic Boundary

The boundary between the Precambrian and the Phanerozoic eons is a controversial and ambiguous topic for palaeontologists. It has changed many times, following new discoveries and changing criteria. However, recent studies have shown that the negative and short-lived excursion in the carbon-isotope composition of sea water coincides well with this boundary. In turn, this negative excursion is constrained by the last occurrence of some types of small, shelly fossils with calcified skeletons like Namacalathus and Cloudina (Fig. 2.29). It was followed by a great diversity and complexity of organisms at the beginning of the Phanerozoic. However, the fossils of these mineralized-skeleton animals are rarely well preserved, particularly in the very changing environment of the late Precambrian and early Phanerozoic.

A number of researchers studied samples of subsurface rocks extracted by the petroleum well Mukhaizna-11 in South Oman in the 1990s and 2000s (Fig. 2.30). The layers of these rocks showed a large spread of two types of mineralized-skeleton organisms, Namacalathus and Cloudina, that date back to late Precambrian time. These layers are overlain by a bed of volcanic ash. Because volcanic ash usually contains uranium isotopes, which are very effective in determining the age of rocks, the layer of ash was accurately dated to be 541 million years old. Since the fossils completely disappeared just above the layer of volcanic ash, this date was considered globally to mark the boundary between the Precambrian and the Phanerozoic eons.

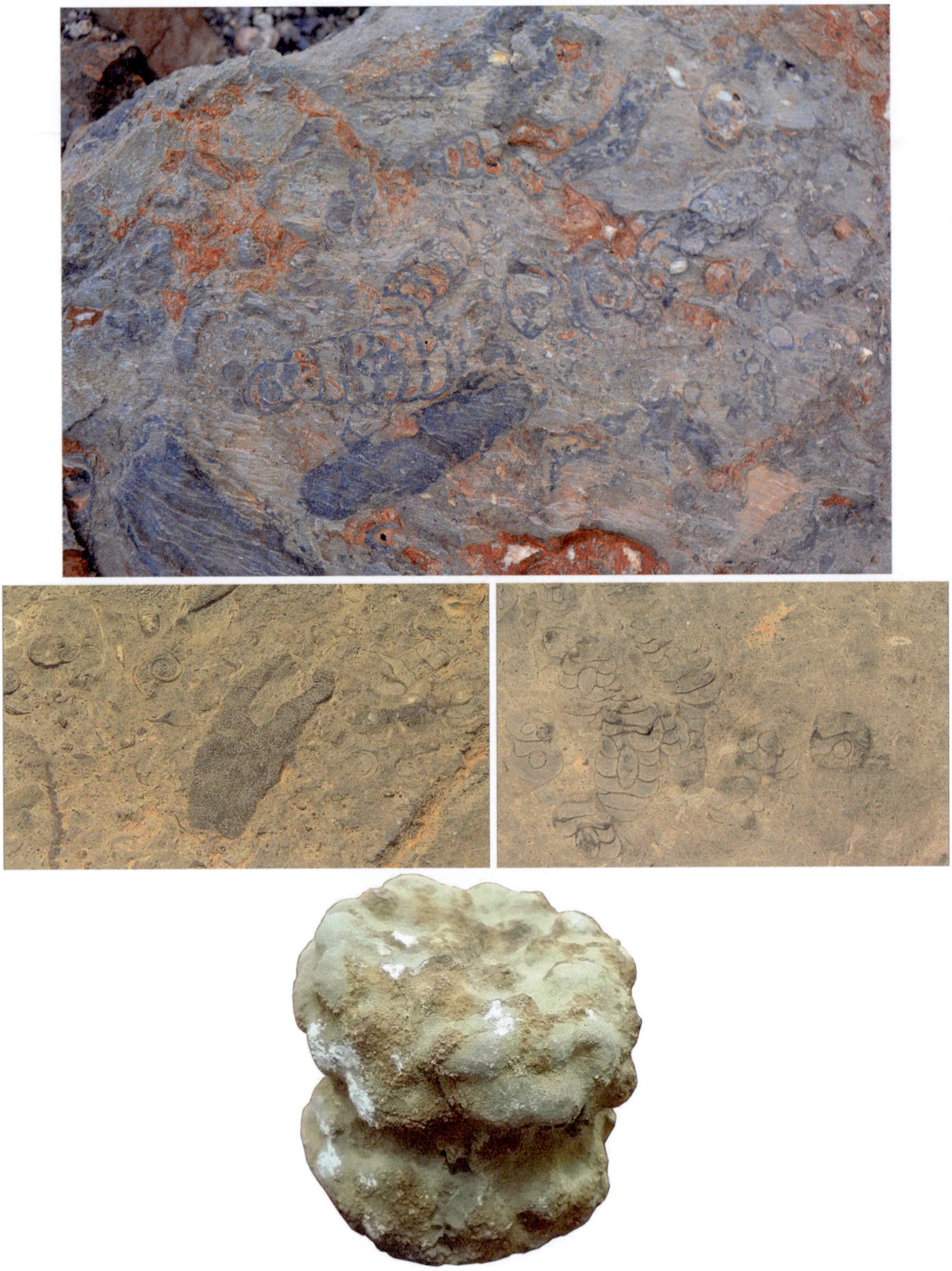

Fig. 2.27 Fossils of different sponges, which represent the most basal animals, each about 10 cm long. These fossils are widespread throughout the Palaeozoic, Mesozoic and Cenozoic rocks of Oman. The pictures on top and in the middle are from the Palaeozoic rocks of the Al Hajar Mountains. The picture on the bottom left is from the Cenozoic rocks of the Dhofar Mountains. Since their appearance in the Precambrian, fossils of sponges occur in all layers of marine rocks in many different forms. They continue to live extensively in the seas off Oman

Fig. 2.28 Living sponges in the shallow seas of Oman are common; in this case, they coexist with sea urchins and corals

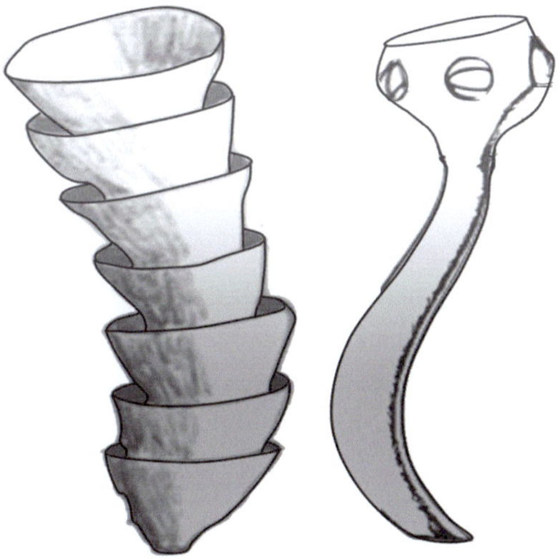

Fig. 2.29 Reconstructions of Namacalathus and Cloudina. Namacalathus is formed of a cup-shaped head and a long, hollowed and curved stalk with tapers at both ends. The stalk could be up to 3 cm long, whereas the cup is only about 0.5 cm long on average, but could reach 2.5 cm in length. The cup is also hollow and has a few holes on the sides. The fossils of Cloudina consist of cones or loops, embedded on top of each other. The size of individual cones could represent growth rates. They possibly grew in microbial mats and were trying to keep pace with sedimentation rates. Namacalathus and Cloudina became extinct at the end of the Precambrian

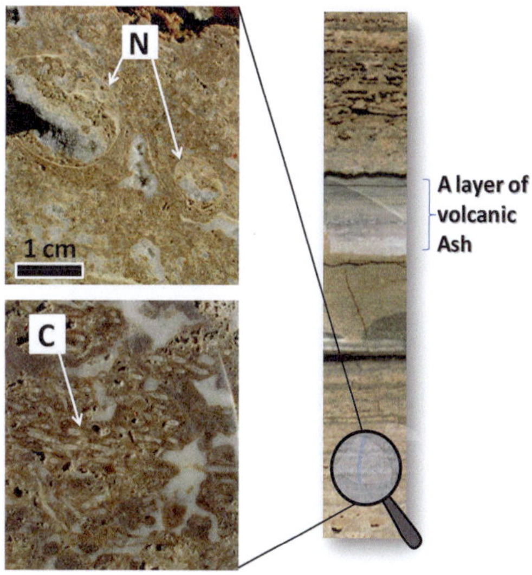

Fig. 2.30 Namacalathus (N) and Cloudina (C) in rock samples extracted from the petroleum well Mukhaiznah-11. The layer of fossils is overlain by a bed of volcanic ash containing isotopes of uranium dated at 541 million years old, as published by Amthor et al. (2003). In these samples (magnified photos), the fossils of Cloudina are found deposited within those of Namacalathus. Frequent volcanic eruptions at that time, along with other environmental changes, led to the extinction of these animals

The genuses of Namacalathus and Cloudina are among the calcified skeletal metazoans (multicellular animals). They are too small to be seen in detail without the use of a magnifying lens or microscope (Fig. 2.30). In comparison with earlier animals, like sponges, the emergence of these skeletal animals was as a big step in animal evolution. Their shells probably evolved from other soft-bodied biota as a protection against predators. Oman and Namibia are among the most famous places in the world where the fossils of Namacalathus and Cloudina are found. They might be primitive corals or sea worms; however, others prefer not to classify them in any animal group. They probably grew in microbial mats—explaining the frequent association of their fossils with stromatolitic structures—and grew vertically at a rate matching the sedimentation rate to keep their structures exposed to sea water.

The rocks of Oman have directly contributed to the provision of a relatively precise date marking the boundary between the Precambrian and the Phanerozoic. Through the fossils of Namacalathus and Cloudina, one of the most debatable scientific topics has found an answer. However, the Precambrian rocks of Oman still lack an important Precambrian biota known as Ediacara, a soft-bodied biota that has fossil records in countries like Australia, England, Namibia, Russia, China and the USA. Finding fossil records of Ediacara remains an open field of search and research in the mountains of Oman. Presence of their fossils in Oman would mean that the land of Oman probably hosted all the known organisms of the Precambrian. It is interesting to note, however, that Cloudina and Ediacara biota normally occur in alternating layers of rocks and not in the same layers.

2.3.5 Summary of Life in Oman During the Precambrian

Figure 2.31 shows the four main stages of development of the land of Oman during the Proterozoic. From left, the first stage represents a hot earth with volcanos that formed during the accretion of the Arabian Plate from smaller microplates around 850 million years ago. During this period, significant volcanic activities occurred particularly along the suture zones between plates. Volcanic dykes (or sometime spelled dikes) form when molten rocks intrude through pre-existing volcanic rocks. These igneous ridges or walls, often

Fig. 2.31 Summary of climates and life forms in Oman during the Proterozoic from the accretion of Arabia under severe volcanic activities, on the left, to the development of various forms of life at the end of Precambrian, on the right

vertical, can be clearly seen along the eastern coast of Oman, particularly in Jabal Qahwan in the Wilayat of Jalaan Bani Bu Hasan or in the Mirbat Plain in South Oman. Sometimes, these dikes consist of a large swarm of smaller dikes that were intruded contemporaneously.

The second stage represents the development of colder crust around 700 million years ago after or during the last phases of continental accretions in the interglacial periods of the Proterozoic. Life was then mostly represented by thin bacterial mats that lived in shallow water which covered many parts of the landmass of Oman. These bacterial mats are formed in carbonate beds which are often overlain by glacial deposits during what is globally known as the snowball earth period. Such mats are found in the Mirbat Plain above the basement igneous rocks.

Later in the Proterozoic, phase 3, Oman was covered by glacial sheets that resulted in the deposition of thick glacial sedimentary layers which can be seen in Al Hajar Mountains in North Oman as well as the Dhofar Mountains in South Oman. Despite the extreme climatic conditions during these glacial periods, life in the form of photosynthetic bacteria and algae has managed to survive, most likely in narrow and deep bodies of water where hot magmatic ridges of oceanic spreading centres exist, like the Red Sea. These organisms formed a bridge for life before and after the different periods of the snowball earth in the Proterozoic.

As the ice melted at the end of the Proterozoic, stage 4 on the right side of Fig. 2.31, shallow water environments were flourished. Stromatolites, microscopic acritarchs, sponges, Cloudina and Namaclathus lived in these shallow seas as illustrated in the figure. The biological domal structures of stromatolites grew to large sizes and form various morphologies depending on the environment in which they grew. Domal, columnar, branching and stratiform shapes of stromatolite fossils from the Precambrian are widely spread in the surface outcrops and subsurface rocks of Oman. The shallow seas of the Precambrian in Oman also included many forms of microorganisms, including different types of acritarchs, as well as sponges, Cloudina and Namacalathus.

3 The Palaeozoic Rocks and Life in Oman

The appearance of plants on land was one of the most important life features of the Palaeozoic. The fossils of plants, their roots, trunks, spores and leaves in Oman have significantly contributed to the understanding of plant evolution in the world during the Palaeozoic. This fossil leaf date back to around 270 million years ago and is from the Al Huqf area of central Oman.

M. H. Al Kindi, *Evolution of Land and Life in Oman: An 800 Million Year Story*,
https://doi.org/10.1007/978-3-319-60152-6_3

3.1 Earth During the Palaeozoic

The Palaeozoic is the first era of the Phanerozoic Eon, which also includes the Mesozoic and Cenozoic Eras. The Palaeozoic extends from the end of the Precambrian, 541 million years ago, to the beginning of the Mesozoic, 252 million years ago. It spans a time period of 290 million years, thus representing the longest era of the Phanerozoic Eon. It includes six different periods (Fig. 3.1); the first three are often known together as the Early Palaeozoic, which consists of the Cambrian, Ordovician and Silurian eras. All of these names are derived from Wales, a country in the UK, with the Latin names of Wales (Cambria) and the principal Celtic tribes living in Wales during the Roman period, the Ordovices and Silures. The later periods, Devonian, Carboniferous and Permian, are subsequently named after Devonshire, a county of England, Latin carbo-fero, referring to the coal-rich beds of this period in England, and Permia or Perm, an ancient kingdom in Russia.

During the Palaeozoic, the earth experienced enormous geological and climatic changes. Life became widely diverse and significantly spread across the land and over the seas. This period started with what has come to be called the 'Cambrian explosion'—a relatively short period in evolutionary terms, lasting a mere twenty million years—when there was such phenomenal growth in terms of rapidity and variety of life forms that most of the phyla known to man first appeared on earth. Interestingly, the Palaeozoic ended just as dramatically with the biggest, most catastrophic mass extinction event in earth's history, when more than 90% of marine species and about 70% of terrestrial species died out completely. Those species that survived took millions of years to recover.

The burst of life at the beginning of the Palaeozoic has been a debatable scientific subject since the mid-nineteenth century. Charles Darwin considered this explosion of life as one of the main arguments against the theory of evolution. The simple question is why did many phyla of the animal kingdom evolve in a period of around twenty million years at the beginning of the Cambrian Period? Perhaps, one of the main reasons for this boom is the increase in oxygen concentrations both in the atmosphere and in water (see Fig. 1.13). Without this increase, life forms would not have evolved and become more complex. Many metabolic functions are impossible without oxygen, and most connective protein tissues, such as collagen, cannot survive in environments lacking oxygen. The ozone layer that absorbs most of the sun's ultraviolet radiation was probably formed just before the Cambrian explosion. Another important factor in the burgeoning of life during the Early Palaeozoic is the increase in calcium concentration in sea water, which allowed animals to build shells and skeletons. By then, the earth no longer had harsh, extreme cycles of warming and cooling as it had during the Proterozoic glaciations. The continental crusts were widely distributed across the planet, enabling complex organisms to diversify in different environments. After 88% of its life, the earth was primed to receive complex life forms and was now quite different from its neighbouring celestial bodies. It should also be mentioned that the apparent burst of life in the Cambrian possibly has a lot to do with many animals acquiring hard parts and so being able to be fossilized, which was not the case with many soft-body creatures in the Precambrian.

Amidst its oceans, during this Eon, a super-continent arose. Known as Gondwana or Gondwanaland, it was formed by the merging of tectonic plates. The Arabian plate was joined with the African and South American plates to form West Gondwana. From the Ordovician period, most parts of this super-plate were positioned over the South Pole, and the whole region experienced an extremely cold climate. The other main continental plates, including Laurentia (Greenland and eastern North America), Baltica (North Europe and Russia), China and Australia, remained close to the equator and were often submerged in temperate water. During the Silurian and Devonian periods, Baltica and Laurentia slowly merged together and, along with parts of Gondwana, surged northward to more tropical climes. Great tracts of warm shallow sea formed along their margins, later becoming perfect environments for marine life to flourish.

The Palaeozoic Era started with a moderate climate. However, by the end of the Cambrian period, sea levels and global temperatures had risen significantly. As plants emerged in the early and middle Palaeozoic, oxygen levels increased, CO_2 levels dropped, and temperatures again rose, but only slightly and with little impact on the diversity or distribution of life forms. By the end of the Palaeozoic, the largest ever supercontinent in the history of earth was formed, when all the continents on the planet merged into one giant continent known as Pangaea. In the centre lay, a vast and arid land exposed to elevated temperatures.

The Palaeozoic Era represents an acceleration in the emergence of animals and plants alike (Fig. 3.2). During this period, nearly all kinds of invertebrates appeared. This era also marks the first appearance of vertebrates, represented by the emergence of primitive fishes that later evolved into larger and more complicated types. Animals began to live on land, possibly with the evolution of tetrapods (four footed animals) and amphibians colonized the shores. By the end of the Palaeozoic (during the Permian Period), reptiles had become quite diverse and spread out across the land. After being purely aquatic during the Early Palaeozoic and possibly the Late Precambrian, plants began to adapt to life on land, too. Starting as algal forms during the Ordovician, by the Carboniferous, they had evolved into complex and gigantic trees. Towering tropical forests grew across vast tracts of dry land Pangaea.

3.2 Oman During the Palaeozoic

The land of Oman witnessed significant climatic changes during the Palaeozoic, resulting in the deposition of different marine and terrestrial rock units. As the eastern and western parts of Gondwana sutured or merged into one supercontinent during the early Palaeozoic, Oman was positioned along the suture or collision zone on the margin of the Arabian and African Plates and was uplifted above sea level (Fig. 3.3). This tectonic event is often known as the Angudan Event which divided the earth's continents into Gondwana and its smaller northern part, Laurasia.

With the collisions of both parts of Gondwana, eastern and western, and the emergence of the surface of Oman above sea level, very thick units of sandstone were deposited across its land (Fig. 3.4). These layers of continental deposits can be clearly seen in the area of Al Huqf in the central part of Oman. They are often inter-bedded with carbonate layers of rock deposited in shallow marine environments as sea levels fluctuated wildly throughout the Palaeozoic. As the sea was filled with diverse living organisms, these carbonate beds are full of different fossil types. Fossils of arthropods, shells and fish are found, together with trace fossils made by these animals as they moved inside or on top of the sand layers, looking for food or hiding from predators. Also, well

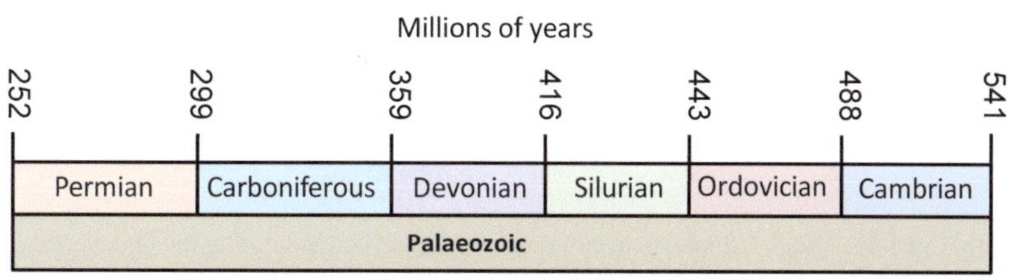

Fig. 3.1 Time periods of the Palaeozoic Era. The Carboniferous Period is currently separated into two periods, the Mississippian (Early Carboniferous) and the Pennsylvanian (Late Carboniferous)

preserved are ripples, which indicate the agitation by the currents or waves of the shallow seas of the Palaeozoic.

The alternation of marine and terrestrial deposits in Oman during the Palaeozoic has preserved both the sea and land organisms of the Palaeozoic. Throughout the Palaeozoic, Oman raced with its mother supercontinent, the Gondwanaland, from a position near the equator to the southern hemisphere near the South Pole (close to the mid-southern latitudes of 40–50°, refer to Fig. 1.5). Oman was exposed to extreme glacial temperatures that left clear grooves and striations, which were caused by clasts embedded within the ice sheet and acted as graving or scouring tools on the surface beneath as they move along with the ice sheets (Figs. 3.5 and 3.6). They are clearly visible in areas like Al Huqf in Central Oman. The grooves are aligned towards the direction of movement of the ice sheets. Gondwana then moved northward, away from the South Pole, to warmer climes. The climate improved and temperatures rose, turning the glaciers into lakes and flowing rivers (Figs. 3.7 and 3.8). On the banks of these rivers, giant trees grew, some of which are more than 15 m high. Oman continued its northerly progress away from the South Pole until the end of this era, about 250 million years ago (see Fig. 1.5).

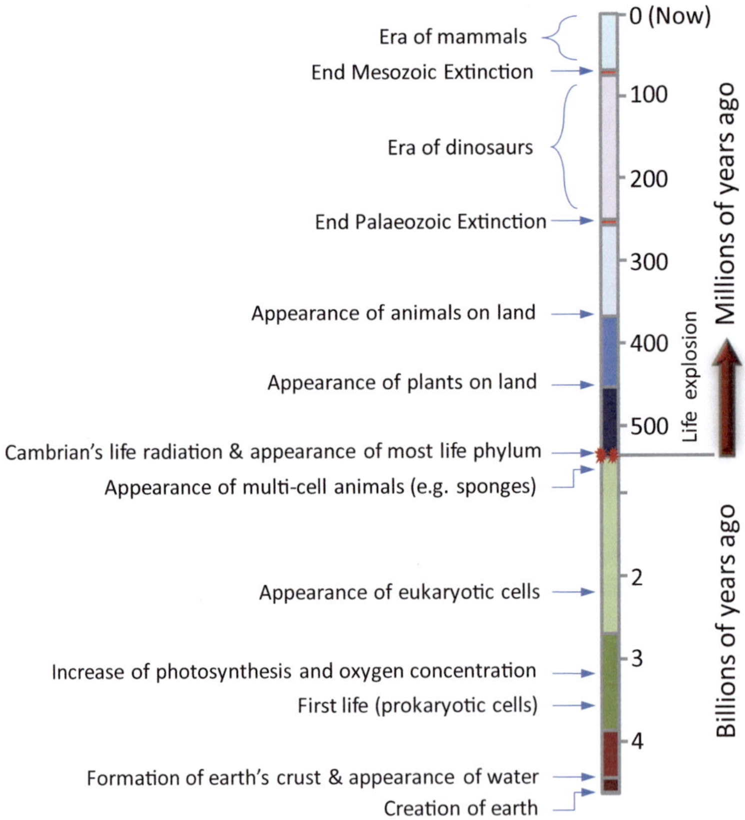

Fig. 3.2 Evolution of life through billions and millions of years, and the explosion of life in the Cambrian Period. The Palaeozoic marks the appearance of most phyla in the animal kingdom. In the Early Palaeozoic, animals like trilobites, brachiopods and molluscs were the primary ones and in the Late Palaeozoic the seas were dominated by various echinoderms, brachiopods and corals. Plants appeared on land during the Ordovician Period and by the end of the Palaeozoic they had colonized vast tracts of land. Invertebrates and vertebrates also became common on land in the Late Palaeozoic. However, life during the Palaeozoic was affected by three mass extinction events that led to the almost complete disappearance of both marine and land species of animals

Fig. 3.3 Suturing of East and West Gondwana during the Early Palaeozoic, about 520 million years ago. This collision event is widely known as the Angudan Tectonic Event. Oman was located along the collision zone (red line), on the eastern side of the Arabian Plate, and thus exposed to a lot of tectonic processes

At the end of the Palaeozoic, Gondwana merged with Laurentia to form Pangaea, the largest ever supercontinent found on earth (Fig. 3.9). In Pangaea, Oman was positioned on the south-western margin of a large ocean known as the Palaeo-tethys Ocean. This ocean formed the inner part of a rather ring-shaped Pangaea.

Because of its strategic position at the edge, the Palaeo-tethys Ocean, Oman, was subject to many different climates and tectonic forces. During the Permian period at the end of the Palaeozoic, massive subsidence of its north-eastern margin was subsequent to the development of a major new mid-oceanic ridge. The land was submerged under the sea of the Neo-tethys Ocean (Fig. 3.10). As the Neo-tethys Ocean was developing and spreading offshore the northern edge of Oman and the Arabian Plate, the remaining part of the Palaeo-tethys Ocean was pushed down under the eastern and northern ends of Pangaea. The Palaeo-tethys completely closed later in the early Mesozoic. In contrast, the Neo-tethys Ocean continued to grow wider with time, at least for a period of about 150 million years, to the end of the Mesozoic, reaching a massive extent from latitude 30° N to the equator. The Neo-tethys Ocean was teeming with marine organisms adapted to live in the shallow seas along the coastal margins of continents, and fossils of these animals are scattered throughout the wadis and mountain peaks of Al Jabal Al Akhdar, the Saiq Plateau and Jabal Shams in Al Dakhiliyah Governorate. Evidence of the mid-spreading ridge that developed at the end of the Palaeozoic in the Neo-tethys Ocean is visible in many different areas in North Oman, one of which is Wadi Al Wasit (Figs. 3.11, 3.12 and 3.13).

The development of the Neo-tethys Ocean over millions of years and the subsequent emplacement of its rocks over Oman at the end of the Mesozoic probably represents the most important geological event that characterizes the geology of Oman. A great portion of the rocks in the Oman Mountains represents the oceanic crust and the oceanic deep and slope sediments of the Neo-tethys.

The Permian rocks of Oman also bear witness to the mass extinction event at the end of the Palaeozoic—the worst ever extinction in the history of life (Fig. 3.14). Its drastic effects can be seen in a number of places in Oman, including Al Jabal Al Akhdar and Wadi Al Wasit.

3.3 Organisms Living in Oman During the Palaeozoic

3.3.1 Trilobites

Trilobites are one of the oldest complex life forms, inhabiting the earth after the dominance of single-celled organisms. Appearing during the Cambrian explosion of life and becoming extinct during the Permian mass extinction, the existence of trilobites covered the whole Palaeozoic Era. Throughout this period, about seventeen thousand species of trilobite of different shapes and sizes emerged, representing one of the most extensive fossil records in existence. Trilobites are marine arthropods which lived in shallow seas. The study of their species has helped scientists understand many details about the

Fig. 3.4 Early Palaeozoic outcrops. Top: Wadi Daiqa in northern Oman. Bottom: The Al Huqf area in Central Oman. Because these sediments often represent alternation between marine and terrestrial deposits, they appear in bright mixed colours

climatic changes, tectonic events and processes of life evolution during the Palaeozoic. They serve as very good index fossils of the Palaeozoic. Their name derives from the Greek trílob (os), meaning three lobed, describing the three skeleton parts that make up a trilobite's body—the head, thorax (body) and the tail. Trilobites were the most prominent organisms in the Palaeozoic. Their eyes, with a great number of lenses, were probably the first compound eyes on earth (Fig. 3.15). The convexity of this eye possibly allowed some trilobite species to see in all directions, giving them superiority against predators and other organisms in competition for food. Some trilobites, however, adapted to living in deep sea where no light penetrated, were probably blind. The segmented bodies of trilobites provided them with flexibility and the ability to contract or curve when required. The tail or pygidium is also curved and usually formed of three segmented parts, one central and two flanking. Trilobites often had a pair of frontal antennae although these are rarely preserved.

The numbers of trilobites significantly decreased with the proliferation of competing organisms and predators in the seas, such as fish in the middle part of the Palaeozoic. Some species or fauna also became extinct in the Early

Fig. 3.5 Straight grooves and striations made by the ice sheets about 320 million years ago in the Governorate of Al Wusta, Al Huqf area, in Wadi Al Khlata South and in Ain Hindi. The striations are oriented NE-SW, which indicates the direction of the ice flow

Fig. 3.6 Boulders in the Al Huqf area of Central Oman with striations on their surfaces. These were made by glaciers that transported boulders and gravel across Oman around 320 million years ago, when Oman was close to the South Pole. The left boulder is 40 cm long, and the right one is about 1 m long

Fig. 3.7 This section in Wadi Al Khlata (Al Huqf) represents a transition from grey glacial deposits in the lower part of the section to fluvial deposits. The flow of the glaciers and fluvials was most likely from south (right) to north (left)

Fig. 3.8 Gharif Pinnacle in the Wilayat of Mahut is formed from sediments that were deposited in the Late Palaeozoic. It marks the stacking of various fluvial channels that flowed from SE to the NW, following the movement of Oman towards the equator to warmer climates and the deglaciation of the ice caps

Palaeozoic (intra-Palaeozoic) extinction phases, but trilobites completely disappeared during the great extinction at the end of the Palaeozoic, although their fossils remain one of the most prominent signs of complex life forms in this era.

Fossils of trilobites have been found in different places across Oman, such as the Wilayat of Jaalan Bani Bu Hasan, Jaalan Bani Bu Ali, Muhut, Qurayyat and Al Amrat. Their fossil record goes back to more than 500 million years to the Cambrian Period and extend to the end of the Permian Period, about 252 million years ago. Therefore, their fossils in Oman span the whole Palaeozoic, for a period of more than 250 million years, and new species are always being discovered (Fig. 3.16). Cambrian trilobite fossils are found in the Governorate of Al Wusta, whereas Ordovician ones are mainly found in places like Qurayyat and Al Amrat. Younger fossils of the Permian Period are found in the Governorate of Ash Sharqiyah South. They can also be found in Al Jabal Al Akhdar, although they are rare. The

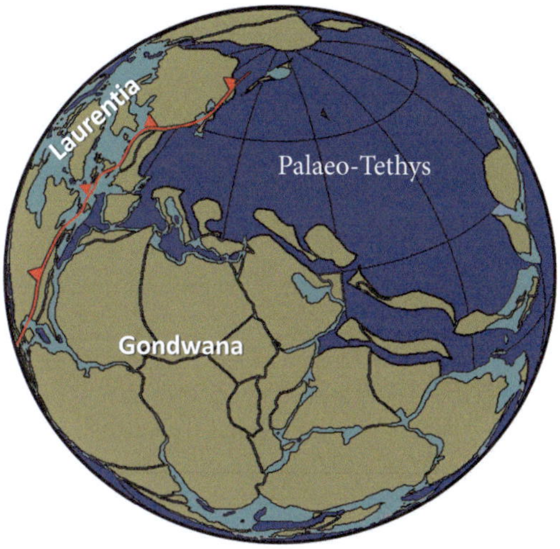

Fig. 3.9 Collision between the supercontinents of Gondwana and Laurentia, 320 million years ago formed the largest ever supercontinent of Pangaea. The collision zone appears as a red line. Note that Oman and the Arabian Plate were positioned away from the collision zone. Oman was then located on the edge of massive ocean, known as the Tethys. This massive ocean was closed at the end of the Mesozoic during the opening of the Indian and Atlantic oceans. The old Tethys Ocean is known as the Palaeo-tethys. This vast ocean was later overprinted by the Neo-tethys Ocean that developed off the shore of Oman

trilobite response to danger is to curl up into a ball, like modern-day armadillos or porcupines. The head meets the tail to protect the vulnerable underside of the animal. This is often the posture in death, too, so a large number of trilobite fossils are found curled or folded over.

Some species of trilobite were great swimmers and ranged across the oceans, whereas others only lived in localized areas on the seabed. As they moved on the sea floor, their tracks left behind grooves and straight lines following the direction of movement (Fig. 3.17). There are at least two main types of trace fossils of trilobites found in Oman: Cruziana and Rusophycus (Fig. 3.18). The latter are probably resting traces as they do not represent any forward movement. Such trace fossils can be found in Wadi Daiqa in Qurayyat or in the Wilayat of Al Amrat.

Fossilized ripple marks that date back to more than 450 million years to the Ordovician Period and were formed by the movement of water over unconsolidated sediments represent the shallow seas environments in which the trilobites lived. These marks can be seen in Al Amrat, Qurayyat and Muhut (Fig. 3.19). Various types of trace fossils are often seen within them. Including tube-shape marks that form between the ripple marks probably represents burrows of Ordovician animals.

3.3.2 Primitive Fish

The first primitive fish appeared around the world during the Ordovician Period, about 460 million years ago. They represent some of the oldest vertebrates that ever existed (Fig. 3.20). During the Early Palaeozoic, they lived in vast, shallow seas on the coastal margins of Gondwanaland. As a result, their fossils are found in Australia and in countries like Bolivia and Argentina in South America. Recently, fossils of primitive fish were also found in the Ordovician rocks of Wadi Daiqa in Qurayyat and in Al Amrat. The discovery demonstrates that these fish were also present in the northern near-shore areas of Gondwana—where Oman was located in the Early Palaeozoic (see Fig. 3.3)—not only at the southern edges of the continent, as previously thought. It has deepened the global understanding of the palaeo-environmental setting during the Ordovician and prompted further sampling and analysis of fish fossils around the world for other potential discoveries not only in South America and Australia.

The recently discovered fish fossils in Oman are an Ordovician species known as Sacabambaspis, after the village of Sacabambaspis in Bolivia where the first fossils of this genus were found. Sacabambaspis lived with trilobites, crinoids and corals in the shallow seas of the Palaeozoic. They had a tadpole shape with a total length of about 25 cm. Their bodies had relatively big heads, long, flexible tails and no clear fins or jaws (Fig. 3.21). The eyes were positioned on the frontal part of the head, like head lamps, but they were probably able to see in different directions. It had relatively large dorsal and ventral webs. Sacabambaspis perhaps fed on

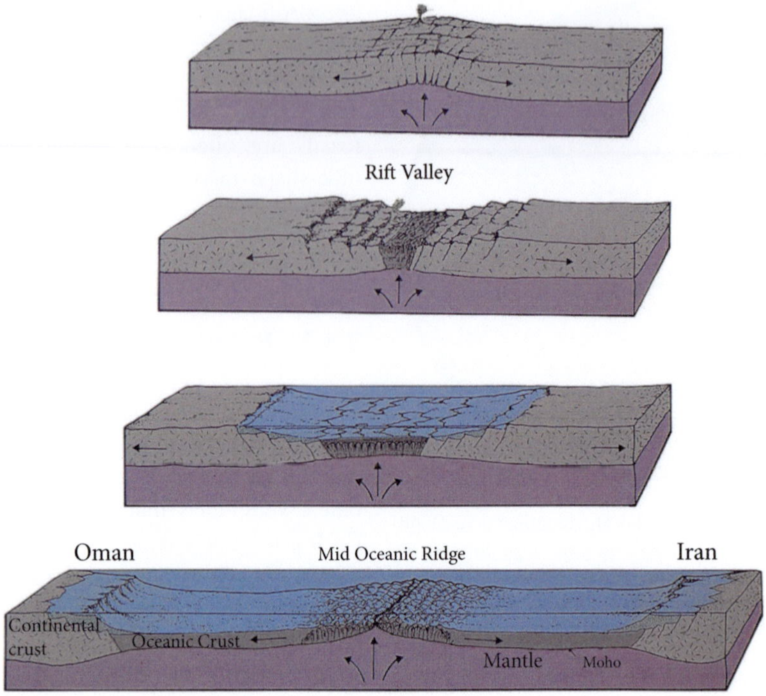

Fig. 3.10 Developmental stages from rifting to the formation of mid-oceanic ridges, similar to what occurred in Oman during the Permian Period as the Neo-tethys Ocean formed. The initiation of the ridge and the formation of the new ocean resulted from mantle upwelling in response to plate tectonics. Molten magma emerges from the centre of the ridge to form a new oceanic crust. The age of the oceanic rocks (Ophiolite) is older away from the ridge than the rocks within the centre of the ridge (Hanna 1996)

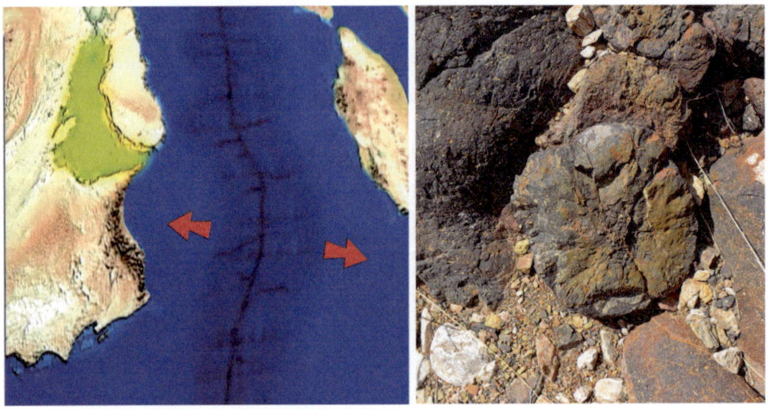

Fig. 3.11 On the left, the formation of the Neo-tethys Ocean at the end of the Palaeozoic Era. On the right, a pillow lava from the middle Permian in Wadi Al Wasit in the Wilayat of Dama Wa At Taiyyin. These pillows are among the oldest known pillow lavas that mark the opening of the Neo-tethys Ocean in the Permian Period

Fig. 3.12 This view from Wadi Wasit in the Wilayat of Dama Wa At Taiyyin consists of a volcanic sequence (dark in colour) made mainly of pillow lavas and tuffites (a mixture of tuff and detrital materials), red shales and the radiolarian chert of deep oceanic sediments, slope sediments with gravity flow deposits and reef limestone. Together these different sequences represent the opening and development of the Neo-tethys Ocean during the Permian and Triassic periods. The fossil-rich reefal limestone is comprised of various types of crinoids, corals, sponges and ammonites

Fig. 3.13 A comparison between rocks formed within the Neo-tethys Ocean (left) and the fossil-rich rocks deposited over the continental platform of Oman (right). At the base of the left photograph are pillow lava of the mid-oceanic ridge (green arrow) and red radiolarian chert of the oceanic sediments and on top are reef limestone blocks (blue arrow). On the right, very fossil-rich shallow marine Permian limestones deposited after Oman's subsidence below sea level

algae and plankton. By the middle of the Palaeozoic, it had evolved to other forms. Their fossils in Oman are mainly scattered pieces. No complete fossils of this early fish have been found in Oman yet. However, there is no doubt that the field is still open to explorers as the samples found in Oman were discovered by accident while analysing and dissolving rock

Fig. 3.14 Impression of the great extinction at the end of Palaeozoic, at the boundary between the Permian and Triassic periods, can be seen as a line (black line in the image) separating dark- and light-coloured rocks in the northern part of the Saiq Plateau on Al Jabal Al Akhdar. The dark-coloured rocks are rich in marine fossils from the Permian Period. These fossils almost entirely disappeared at the boundary line, before gradually reappearing in the lighter-coloured rocks, although occurring at much lower concentrations and of differing types

samples in search for an organism known as Conodont, a primitive eel.

3.3.3 Nautiloids

Nautiloids are marine molluscs that belong to the Nautiloidea subclass and cephalopod class. This class is represented today by a limited number of exclusively marine animals, such as the octopus, squid and nautilus. Unlike the octopus, the nautilus does not have the ability to produce ink as a defence. Instead, it depends entirely on its outer shell to protect itself from predators. Cephalopods were very common during the Palaeozoic and Mesozoic eras, and they probably first appeared during the Cambrian Period as primitive nautiloids. However, their numbers were significantly reduced during the Cenozoic. They are formed of a twisted or coiled shell divided into camerae or chambers which grow in number and size as the animal matures. The prominent head and tentacles are positioned in the largest chamber at the end of the shell. The animal could completely hide inside and close its opening when faced with danger. The straight to sinuous suture lines on nautilus shells can be used to determine the age of the nautilus, a little like counting growth rings to find the age of a tree. The shells of dead nautiloids are often used by people as an artistic ornaments or jewellery because of their beautiful shape and colouration. Their fossils are dominant among cephalopods and are probably the most well known. Unlike many creatures of the Palaeozoic, there are extant species of nautiloids. They normally live at depths around one hundred metres, although some—in the Pacific and Indian Oceans—are found as far down as 700 m. They are tremendous swimmers, moving by jet propulsion using their powerful muscles, and in fact, their name stems from nautilus which means 'sailor' in Greek. The widespread distribution of fossil forms suggests they are aptly named.

Two main types of nautilus fossils are found in Oman: the orthocone, which has a long straight shell, and the nautilus, which has a coiled shell (Fig. 3.22).

3.3.3.1 Orthocones

Unlike the coiled nautilus that can be found today in deep waters around the world, orthocone

Fig. 3.15 Various trilobite fossils found in Oman. Parts of the head and the tail of different types of trilobite found in Ras al Had and the Wilayat of Al Amrat, where many types of trilobites lived. The Ras al Had samples all belong to the Permian Period and occur along with a diverse fauna of marine molluscs. The top sample was found by Alan Heward and could be a new species. The last two samples are Ordovician trilobites from Wadi Al Sareen in Al Amrat. The samples are about 3 cm long

fossils represent the straight, uncoiled shell of an extinct type of early nautilus. Orthocones most likely appeared in the Cambrian Period and became extinct during the Early Mesozoic Era. They were very common in the Palaeozoic, reaching a few metres in length and probably the largest animal on earth during the Early Palaeozoic. As a result of their size and tough shell, they were most likely the biggest predators during the Ordovician and Silurian periods, feeding on vertebrates and arthropods. The mouth and tentacles emerged from the large end of the shell. These tentacles were used to seize prey, as well as to propel themselves through the oceans by forcing the water in the opposite direction. Fossils of orthocones are prevalent in many places in Oman. Well-preserved examples can be found in the Ordovician rocks of Wadi Daiqa and the Permian rocks of Al Jabal Al Akhdar (Fig. 3.23).

3.3.3.2 Nautilus

The nautilus, with its coiled shell, is often known as the living fossil. Although most of its kind became extinct, some species are still extant today—relicts from some 500 million years ago when the first nautilus appeared. They have

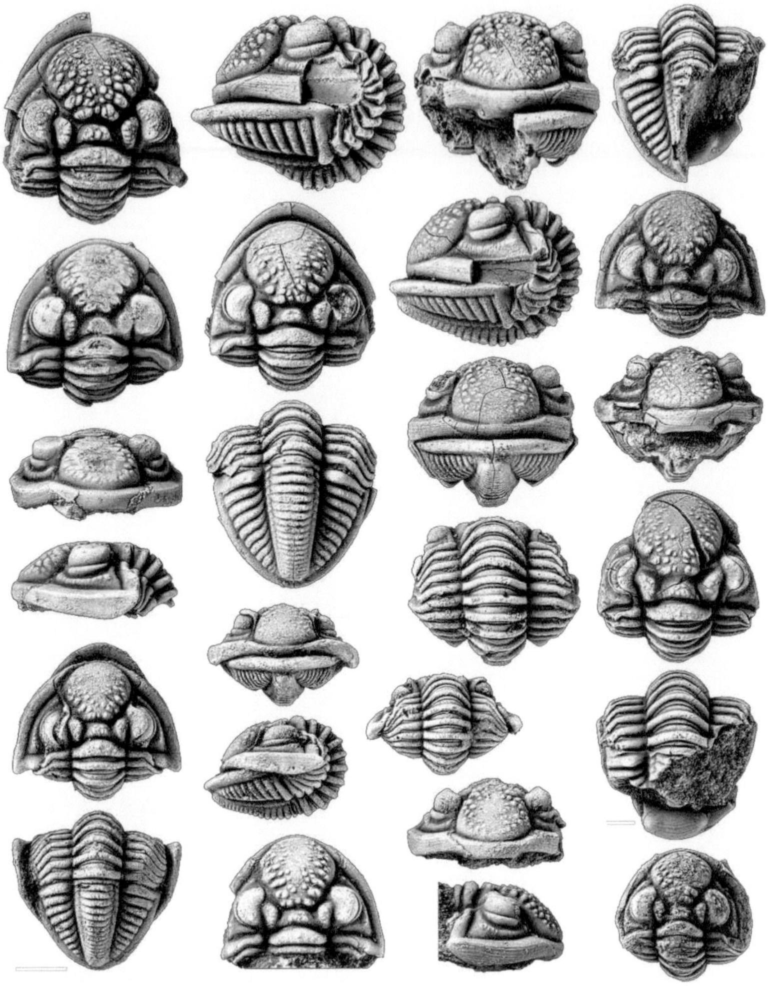

Fig. 3.16 A number of diverse Permian trilobites from Fortey and Heward, 2015, collected from the Ras Al Had area. Overall, the trilobites found in this area include 5 new trilobite species (mainly belonging to the family Phillipsiidae). These fossil samples are considered well preserved compared to other samples from the Permian Period. All the samples in this figure are 1.5–4 cm long

changed very little in all that time. The fossils of nautiluses can be seen in all the rock units of the Phanerozoic Eon, including the Palaeozoic, the Mesozoic and the Cenozoic. Palaeozoic examples can be seen on Al Jabal Al Akhdar and at Al Huqf. Mesozoic and Cenozoic fossils of nautilus can be seen in the Wilayat of As Seeb, where some reach 30 cm in length (Fig. 3.24). However, it is almost impossible to be sure where nautiluses have lived exactly along the vast seas, because their shells may travel long distances. When the animal dies, the soft tissue decomposes and gas becomes trapped inside, allowing the shell to float. As the gas dissipates, the shell sinks to the bottom or is tossed on shore and so may be deposited in a completely different area.

3.3.4 Bivalves

The shells of bivalves are composed of two parts, frequently identical in size and shape, with an interlocking segment which fastens the shells closed. The term comes from Latin words bi, meaning double, and valva, meaning one half of a folding door. Bivalves are a class of fresh and marine water molluscs and include numerous families such as oysters and clams (Fig. 3.25). They represent an essential element in the ecological system of salt and freshwater. They first emerged during the Cambrian Period, more than 500 million years ago, although many species have since become extinct—mainly during the mass extinction events. Their fossils are very

Fig. 3.17 Slab of Cruziana traces in the Wilayat of Muhut. The whole rock slab is about 1 m long. One of the Cruziana traces in Muhut is named C. Omanica, and it characteristically has 3 claws with the central one deepest

Fig. 3.18 Figure on top shows how trilobites moved across the sea floor of shallow seas, from 'The Story of Life in Oman' film. The middle photograph is for Cruziana trace fossils from the Cambrian Period found in the Al Huqf area and preserved in the Natural History Museum in London. The lower photograph represents Cruziana and Rusophycus trace fossils from the Ordovician Period found in Qurayyat and Al Amrat. The trace marks are from the bases of beds, and these are casts of the tracks made by the animal. The widths are 6–12 cm

common and conserved in many Phanerozoic rock units. This is primarily because of their hard and resistive shells composed of calcium carbonates, which make bivalves (and many molluscs in general) one of the best-preserved animals that have ever lived. Inside the shell is a powerful foot that helps the animal to move and dig the sediment. It also includes a digestive system, nerves, muscles and excretory systems.

Bivalves have been used as a food source by humans for thousands of years, as well as providing decoration and being used as cutting tools by early man. In addition, the study of bivalve shells has provided a lot of information about life's evolution and climatic changes throughout the history of earth. They diversified widely and

Fig. 3.19 Fossils of ripple marks formed by the movement of shallow seas during the Ordovician. Small rounded marks can be seen between the ripples. These might be burrows created by Ordovician animals

Fig. 3.20 Reconstruction of Sacabambaspis, a fish that lived in shallow water during the Ordovician Period, about 450 million years ago. This armoured jawless fish has a total length of around 25 cm and has a flat body, big head and front forward eyes. The fish had a head shield and a body covered by long scales. Complete or fragmentary fossils of this fish were found in Bolivia, Central Australia, Argentina and more recently in Oman. The discovery of Sacabambaspis in Oman is important because it shows that this fish lived around the northern and southern margins of the Gondwana palaeocontinent and not only in the southern margins, as previously thought. Therefore, it indicates potential future discoveries in similar rocks from the same age in other parts of the world

spread extensively during the Ordovician Period. By the Silurian, they were the most common filter feeders in marine environments. However, many species of bivalves became extinct at the end of Palaeozoic, and new species appeared in the Mesozoic.

Fossils of various species of bivalves are distributed throughout a number of rocks units of the Palaeozoic. Although Palaeozoic bivalves are less common compared to the Mesozoic and Cenozoic ones, they dominated specific habitats and therefore look abundant in certain rock units. New species of Permian bivalves have been identified recently from the outcrops and subsurface rocks of the Arabian Gulf and Oman. They indicate that bivalves have significantly evolved during the Permian Period before suffering from the end-Permian extinction. Only two examples of Palaeozoic bivalves are presented here, ridonia from the Early Palaeozoic and alatoconchid from the Late Palaeozoic (Fig. 3.26).

3.3.4.1 Ridonia Bivalves

One of the common bivalves in the Early Palaeozoic is ridonia which thrived in the shallow seas. It is one of the oldest bivalve types and

Fig. 3.21 Fossils of Sacabambaspis published by Sansom et al. (2009). The fossil on the left represents the head shield of the fish, whereas the fossil on the right is of a flank scale. Each sample is magnified about 10 times. These specimens are housed at the Natural History Museum, Muscat, Oman. They were found in the Wilayat of Qurayyat in Wadi Daiqa Ordovician rocks. The fossils are about 2 cm long

Fig. 3.22 A drawing and reconstruction of two main members of the Nautiloidea subclass, on the left is the extant nautiloid and on the right is the extinct orthocone. Nautiloids belong to the cephalopods class, which is a branch from the molluscs phylum. Thousands of species of nautiloids have lived since their first appearance in the Palaeozoic, but only few of these still live today

probably lived close to the shore in the brackish water of estuaries, where the freshwater meets the open sea, as indicated by the rocks hosting their fossils. Redonia was a simple shallow burrowing bivalve and was frequently eroded out by storms and concentrated in shell beds. They could be among the first organisms to adapt to live in relatively freshwater with low saline levels. Their fossils are common in the Ordovician rocks of Qurayyat and Al Amrat (Fig. 3.27). These rocks were most likely deposited in the early Ordovician; hence, the fossils of ridonia in Oman could be among the oldest of their type in the world.

3.3.4.2 Alatoconchid Bivalves

Alatoconchid bivalves are a group of Permian bivalves which both appeared and disappeared in the Permian. They are among the largest of Palaeozoic bivalves, reaching sizes of up to a metre across. This is probably an adaptation to living in shallow seas of less than fifty metres deep

Fig. 3.23 Fossils of orthocones. The top picture is from the Ordovician rocks of Wadi Daiqa. The picture in the middle is from the Permian rocks of Al Jabal Al Akhdar. These fossils are well preserved. The possible imprint of tentacles is clear at the end of the shell. The length of each sample is about 10 cm. The fossils clearly show the septa between the different chambers of the animal. On the bottom left, a curved or loosely coiled shell of Nautiloidea from Al Jabal Al Akhdar. These fossils probably represent a transition from the straight-shelled animals to the coiled-shell ones

and in a warm climate. While their size might have been advantageous in terms of protection from predators, it could have as well restricted the movement of these bivalves which probably required a constant rich supply of nutrients. Although shell fragments of around half a metre long are very common in the Permian rocks of Al Jabal Al Akhdar, complete fossils are scarce (Fig. 3.28). Along with Oman, fossils of alatoconchid bivalves can also be seen in countries like Japan, Iran, Afghanistan and Croatia.

3.3.5 Graptolites

Graptolites are extinct animals belonging to the phylum Hemichordata. The name is derived from the Greek graptos and lithos, meaning 'written' and 'rock'. Indeed, their fossils look like ancient writings or drawings in the rocks. Graptolites were very common during the Palaeozoic, from the Upper Cambrian period, but they became extinct at the Lower Carboniferous. They were mostly attached to the sea floor, but some species possibly drifted freely in the water. They were arranged in colonies composed of one or more cup-shaped structures with a number of branches. Each of these branches housed an individual zooid. Because of their worldwide distribution and number of different species which evolved within a relatively short period of time, graptolite fossils are often used as index fossils to date Palaeozoic rocks, particularly Ordovician and Silurian. Geologists in the petroleum industry in Oman, such as Petroleum Development Oman (PDO), might use graptolites—which appear as dark carbonized film in the subsurface cores—to date Palaeozoic rocks (Fig. 3.29). However, Graptolites are not common in Oman, so they are not frequently used to date rocks. They were used in the late 1950s to date the mid-Ordovician interval of the early petroleum wells.

Fig. 3.24 Nautiluses. On top is a picture of a living nautilus (Photograph from Monterey Bay Aquarium), showing its prominent head, many tentacles and developed eye structures. They swim by pumping water in and out of the living chamber. These animals have survived relatively unchanged for millions of years. In the middle is a Permian fossil of nautilus from Al Jabal Al Akhdar. On the bottom right are three Cenozoic nautilus fossils from As Seeb. On the bottom left is another nautilus fossil from As Seeb. The bottom samples are 30 cm long. The nautilus shell is the logo for Petroleum Development Oman, although people argue whether the logo is a nautilus, an ammonite fossil or maybe a type of foraminifera

3.3.6 Corals

Corals are marine invertebrates that had appeared by the Early Palaeozoic. Today, they colonize vast areas of shallow seabed to form great reefs, such as the Great Barrier Reef in northern Australia. Corals belong to the phylum Cnidaria which contains a huge number of species. Many types became extinct during the great mass extinction at the end of the Palaeozoic Era, but the whole phylum is characterized by a body cavity which allows water circulation for respiration and digestion. Corals can be either solitary or colonizing. Solitary corals have a single

Fig. 3.25 Clams—a living form of bivalve on the coastal areas of Oman

Fig. 3.26 Artists' impressions of bivalves. On the left side: top view and side view of ridonia type. On the right side: top view and side view of alatoconchid type

Fig. 3.27 Ordovician rock units containing layers of ridonia in Wadi Daiqa in Qurayyat and the wadis of Al Amrat

horn-shaped stalk known as a polyp. Corals which form colonies form masses or have many branches. They probably first appeared in solitary forms during the Cambrian Period, before managing to form large colonies in the Ordovician. The bodies of corals are enclosed in skeletons formed of aragonite, a form of calcium carbonate mineral. These are open at the top to allow tentacles to extract food and expel waste. However, when danger is faced, the tentacles completely contract inside the skeleton. Coral fossils are widespread among the Palaeozoic rocks of Oman, particularly those exposed in Al Jabal Al Akhdar, where the two main types of corals are rugose and tabulate (Fig. 3.30).

3.3.6.1 Rugose Coral

Rugose corals—also often known as Tetracoralla—can be found in the Palaeozoic rocks of Oman spanning the period from the Ordovician to Late

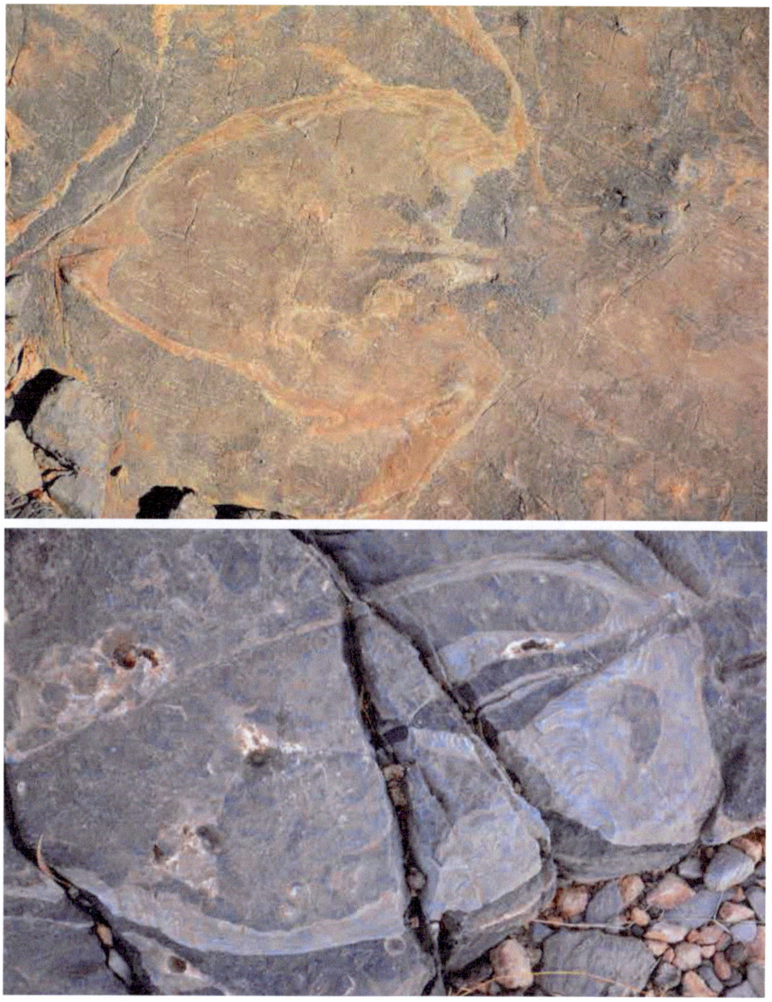

Fig. 3.28 Nearly complete fossils of alatoconchid in the Permian rocks of Al Jabal Al Akhdar. These fossils are about 30 cm long. Some extend for nearly a metre, and they are well preserved in some limestone facies. Fragments of these bivalves are widely spread in the Permian rocks of Al Jabal Al Akhdar, but the most complete fossils can be found close to Al Jabal Al Akhdar Hotel on the Saiq Plateau

Permian. They are found in both solitary and colonial forms. The singular corals or polyps are similar to horns with rugose or wrinkled calcite walls that represent their skeletons. The length of their horns sometimes reached a metre in length. They are often arranged in a reefal framework of several metres in diameter, as seen in the Permian rocks of Al Jabal Al Akhdar (Fig. 3.31). The extensive and scattered spread of rugose coral fossils in Al Jabal Al Akhdar indicates that this area had a perfect tropical climate of a fertile shallow sea in the Permian Period, when Oman was located near the equator.

Permian rugose corals can also be found in Ras Al Had; however, the coral fossils in this area are mainly solitary. Individual polyps of rugose corals are well preserved, but they are most likely less diverse and less widespread compared to the ones on Al Jabal Al Akhdar (Fig. 3.32). This could be related to a deeper sea environment at Ras Al Had during the Permian compared to the shallower environment of Al Jabal Al Akhdar which allowed the corals to spread and grow as large coral colonies. The polyps of rugose corals in Ras Al Had are found with many other Permian fossils of trilobites, crinoids and blastoids. Rugose corals are usually very sensitive to chemical and physical environmental changes, such as the changes in salinity, temperature or sea level. The plates that divide the skeletons of some rugose corals species could be indicative of time periods. These

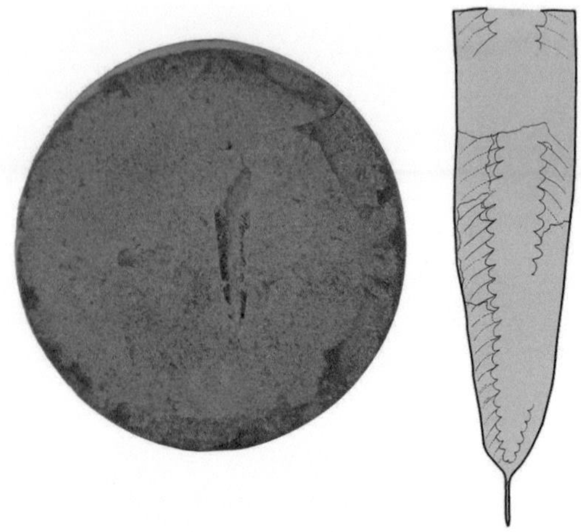

Fig. 3.29 On the right is a drawing showing the shape of a graptolite in Oman Rickards (2007). On the left is a graptolite fossil, Didymograptus, from the Early to Mid-Ordovician, recovered from a petroleum well in Oman. The fossil sample in the middle of the core is about 3 cm long. Each colony of graptolites is about 5 cm long. Graptolites are colonial and benthic (attached to the sea floor) animals. They are used as index fossils to date rocks in the Early Palaeozoic because of their wide diversity and rapid evolution over a relatively short time period. They are also useful to estimate water depths and temperatures in the environment of deposition. Most graptolites became extinct at the end Ordovician mass extinction. Each colony of graptolite has a number of branches known as stipes and it originates from an initial blade or theca known as the sicula

plates or rings grow at certain rates every year, and during the Devonian, the plates grew annually at a rate of 410 loops, compared to the present 365 loops per year. This suggests that there were possibly more days per year in the past compared to now, perhaps due to the slowing speed of earth's rotation around the sun as time passes.

3.3.6.2 Tabulate Corals

Tabulata corals are just as common as Rugosa in the Palaeozoic rocks of Oman. They represent another extinct form that appeared in the Ordovician and died out during the mass extinction at the end of the Palaeozoic (Fig. 3.33).

Tabulate corals are exclusively colonial, forming slender and closely spaced corallite tubes of circular, oval or polygonal shapes. Colonies of tabulate corals are usually smaller than those of rugose corals. There are about 300 identified species, one of which, Multithecopora omaniensis, was named after Oman in 1978, as it was first described here and it is widely found on Al Jabal Al Akhdar.

3.3.7 Echinoderms

Echinodermata is one of the largest phyla of animals on earth. It is a marine phylum, with about 7,000 extant species and about 13,000 extinct ones. None of them live in freshwater. The name is derived from the Greek words echinos, meaning 'hedgehog', and derma, meaning 'skin'. Starfish, sea urchins and sea cucumbers are all echinoderms. Some, like crinoids live attached to the seabed, while others, like sea urchins, move about. They first appeared during the Cambrian explosion, although some organisms from the Precambrian which have been incorrectly classified may belong to this phylum.

Echinoderms live in both shallow and deep sea water. Their skeletons are composed of calcium carbonate plates. They have a number of spines and tube feet that, respectively, protect the animal and pass food towards the central mouth. A number of Palaeozoic echinoderm fossil types are found in Oman, but the three commonest are crinoids, blastoids and echinoids (Fig. 3.34).

3.3.7.1 Crinoids

Crinoids, or sea lilies, are formed of a stem with attached arms, and although they look like plants, they have well-developed digestive and reproductive organs, as well as a mouth, anus and arms that move food towards the mouth. They live in both shallow and deep seas. The skeleton has a number of disc-shaped plates, and these are usually what survive in the fossil record. Crinoids first appeared during the Ordovician and were far more numerous in the Palaeozoic than they are today. A few hundred species survive, but there are thousands which disappeared during the mass extinction at the end of the Palaeozoic.

In Oman, fossils of crinoids are common in many places, particularly in the Palaeozoic and

Fig. 3.30 On the top left, various types of corals close to the Ad Dimaniyyat Islands in the north of Oman. Overall, there are tens of species of corals from offshore Oman, making this country a paradise for coral reef watchers from all over the world. On the top right is a reconstruction of solitary rugose coral (drawing by Lena Cole) and on the bottom left a colony of rugose corals by the Milwaukee Public Museum. On the bottom right, is a reconstruction of Syringoporid tabulate coral, one of the tabulate coral species, modified from Nelson (1977)

Mesozoic rock units. Fossils of crinoids from the Early and Late Palaeozoic can be found in the wilayats of Muhut, Ad Duqm, Sur and Nizwa (Al Jabal Al Akhdar). The oldest crinoid fossils are from the Ordovician period, but these are rare everywhere in the world. Some of the oldest were found in Wadi Daiqa in Oman (Fig. 3.35). Also, recently found in Oman are some new species of Permian crinoids that had not been previously identified (Fig. 3.36).

3.3.7.2 Blastoids

Blastoids are part of the phylum Echinodermata. They first appeared during the Cambrian or Ordovician periods and became widespread in the seas of the Carboniferous before completely disappearing in the mass extinction at the end of the Permian. As a result, fossils of these animals can only be seen in Palaeozoic rocks. They are similar to crinoids, with a body formed of a main theca or the covering case. The theca is attached to a stalk formed of disc-shaped plates. This relatively short stalk—absent in some species—is attached to the sea floor. The mouth was located on the top of the theca and surrounded by thin structures that trapped food and directed it towards the mouth. Formed of interlocking segments of calcium carbonate, the theca is usually the only part of the blastoids to be preserved in rock (Fig. 3.37). Unlike many other animals, the segments of the theca remain attached after death, hence helping to preserve the whole theca. The fossils of many species of blastoids are well preserved near Ras Al Had in the Wilayat of Sur, some of which are rare and have not been found before, except in West Timor. These fossils belong to the Permian period.

3.3.7.3 Echinoidea

Unlike the crinoids and blastoids, echinoids—sea urchins—are not attached to the sea floor. Instead, they swim slowly in sea water and have no attached stalks or tentacles. They possibly appeared at the end of the Ordovician Period and managed to survive the Permian mass extinction. An estimated 950 species of sea urchins still live today in shallow and very deep sea water (As deep as 5 km). Like many other Echinodermata classes, their bodies are composed of five segments that form a very obvious symmetrical shape that characterizes their fossils.

The body of urchins is almost completely covered by spines, therefore protecting the whole

Fig. 3.31 Waageno-phylum (top) and Wentzelella (bottom) rugose coral colonies in the Permian rocks of Al Jabal Al Akhdar. These two types of coral are widely spread in the Permian carbonate rocks of Oman. They form a large part of the fossil coral reefs from that period on Al Jabal Al Akhdar. The area also includes many types of Permian marine fossils that lived in the coral reefs that provided and continue to provide important feeding and nursery areas for fish. Waagenella colonies could exceed a metre in length, whereas those of Wentzelella rarely exceeded half that

animal, but these spines are only rarely preserved with the calcium carbonate exoskeleton (Fig. 3.38). However, their original positions are often well marked. Most Palaeozoic sea urchin fossils are not well preserved, but Mesozoic and Cenozoic fossils are found worldwide and are well preserved in many rock units across Oman, particularly the shallow marine Cenozoic rocks in North and South Oman (Fig. 3.38).

3.3.8 Brachiopods

Animals of the phylum Brachiopoda were among the most prevalent marine animals during the Palaeozoic. They were even more widespread and diverse than bivalve molluscs. However, many brachiopod species became extinct in the mass extinction of the Late Palaeozoic. New species developed during the early Mesozoic, but their development was much slower than bivalves; hence, they became less dominant. Nowadays, there are 330 species of brachiopods still living. The name is from the ancient Greek words brachio, meaning arm, and podes, meaning feet. Its two valves form the body, and it has a pedicle or stalk which attaches it to the sea floor. The two valves are not identical, and they connect to the end, with the upper valve opening or closing for feeding and protection (Fig. 3.39).

Fig. 3.32 Polyps of rugose coral from Ras Al Had (top) and Al Jabal Al Akhdar (bottom). Every corallite is divided vertically by a number of walls or septa. They look like horns; therefore, they are commonly called horn corals, with a cup-shaped top where the polyp lived. The skeleton of the corallite is made of calcite, and this is the part of the animal that gets fossilized. The wrinkled appearance of the skeleton is the reason why these corals are known as rugose corals. Solitary and colonial rugose corals became extinct at the end of the Palaeozoic Era during the mass extinction of the Permian Period. Each horn coral is about 5 cm long

Brachiopods are divided into two main types: articulate (jointed) and inarticulate. The valves of the first type are mainly made of calcium and have a simple muscle system for opening and closing, with toothed hinges to connect the two valves. Inarticulate brachiopods, however, have a more complex system of muscles that keep the two valves aligned, and they are mainly made of calcium carbonate. Overall, the valves of brachiopods have strong muscles that can quickly and firmly close the two valves for long time when danger is faced. Brachiopods usually prefer to live in quiet marine environments, away from strong currents.

The fossils of brachiopods are widely distributed in the Palaeozoic rocks of Oman (Fig. 3.40). They are abundant in Al Huqf area and in Ras Al Had. Their fossils can be distinguished from those of mollusc bivalves, because the two valves in bivalves are symmetrical, whereas the valves of brachiopods are not and they exhibit different line patterns outside. Moreover, the two valves of brachiopods usually remain closed when the animal dies, while the valves of bivalves open.

3.3.9 Trace Fossils of Unknown Organisms

As the forms of complex life proliferated during the beginning of the Palaeozoic, animals made

Fig. 3.33 Tabulate corals of Multithecopora omaniensis species, as named by Oekentorm 1978. The fossil samples of this coral in Al Jabal Al Akhdar are usually 10–30 cm long. Tabulate corals of different species went extinct at the end of the Palaeozoic during the Permian mass extinction event

Fig. 3.34 Three different types of echinoderms. On the left is the general shape of crinoids showing the crown at the top with its arms and cup, and the stalk with its columnar rings and holdfast. In the middle is the overall shape of blastoids with the stalk made of stacked disc-shaped plates and the covering theca on top. On the right is a sea urchin with the spines connected to the skeleton. All extant and extinct species of echinoderms are found only in the sea, in various habitats, including shallow and deep seas, and cold and tropical sea water. They mainly live attached on the sea floor. The bodies of echinoderms are composed of different segments that are radially symmetrical and they have tentacle-like structures attached to them. Fossils of different types of echinoderms are found abundantly in the Palaeozoic, Mesozoic and Cenozoic carbonate rocks of Oman, and many species of echinoderms live today off the coast of Oman

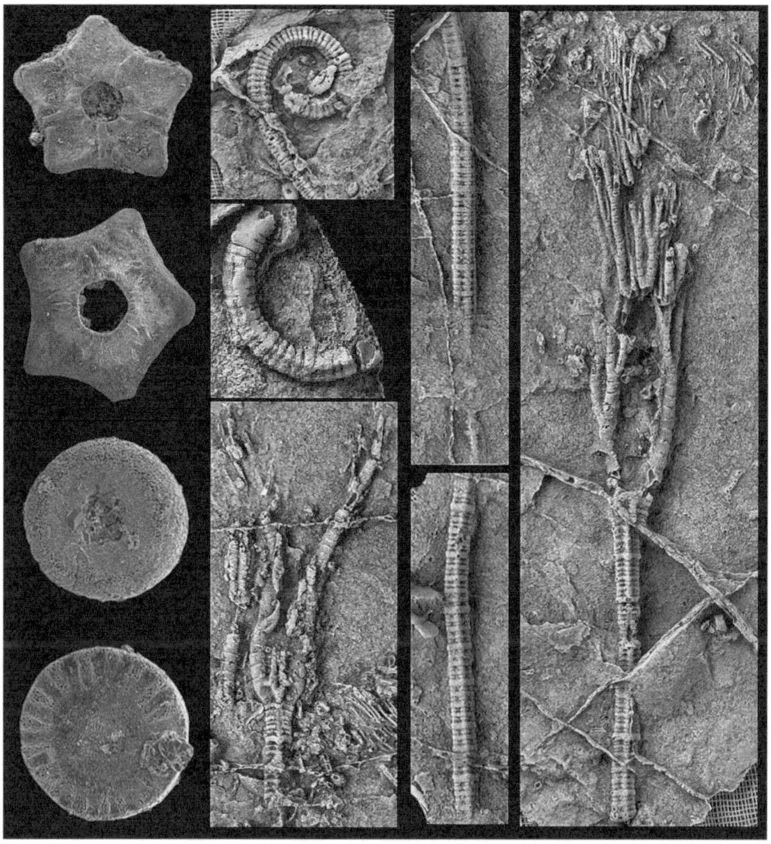

Fig. 3.35 Latex casts of fossils of Ordovician crinoids (around 470 million years old) that were found in Wadi Daiqa and were published by Donovan et al. in 2011. These fossils are considered among the oldest fossils of crinoids in the world. They are characterized by having arms that branch at least seven times instead of four arms as seen in other places in the world. These casts represent longitudinal and cross sections of the samples. The samples are about 15 cm long and 1 cm wide. These Omani specimens are important because they represent the oldest known *Iocrinus* (an extinct genus of Ordovician crinoids) in the world and also the first Middle Ordovician crinoids described in Arabia

many traces in the sediments, recording their presence as fossils which captured their activities as they rested, fed, dwelt and hid. These trace fossils are important in understanding the ancient environments in which the creatures lived and their living and feeding habits. Such fossils have been preserved in many Palaeozoic rocks in Oman (Fig. 3.41). In many cases, the animal responsible for the trace is unknown. Examples can be seen in the northern side of Al Huqf, where burrows of a Cambrian animal that fed on algae were found in shallow sea deposits. Vertical and cylindrical dwelling burrows, known as Skolithos, are abundant in the Ordovician rocks of Qurayyat and Al Amrat. Some are more than 30 cm long, and they were probably made by marine worms that lived in the shallow marine environment of the Ordovician. Other trace fossils, Daedalus, which look similar to a garlic clove are found in the same rock units. They possibly represent the resting or hiding places of Ordovician animals. Additionally, many undefined structures in the Wilayat of Al Amrat could be trace fossils of Early Palaeozoic animals that lived in shallow seas (Fig. 3.42).

3.3.10 Bryozoa

Bryozoa is a phylum of minute animals which first appeared during the Early Ordovician period. No more than 0.5 mm in length, they are hard to see as

Fig. 3.36 Top left: a living crinoid close to the Ad Dimaniyyat Islands in northern Oman. Other pictures show different types of Permian crinoid fossils from Oman, from around 260 million years ago. On the top right is a Permian crinoid that represents a new species of crinoid found in 2002 in Ash Sharqiyah South Governorate by Dr. Alan Heward. The sample is about 10 cm long. Middle-top is a crinoid from the Al Huqf area, which is about 15 cm long. In general, there are many taxa of Permian crinoid in the Al Huqf area, one of which is a species known as Moapacrnus omanensis. In the middle is a collection of various crinoid fragments from Ras Al Had, from 1 to 4 cm long, and at the bottom is a long crinoid stem fossil, about 20 cm long, from Ash Sharqiyah North Governorate

individuals, but usually live in colonies or clones known as zooids. These are quite similar to the colonies of individual coral or polyps, but bryozoa are much smaller and have tiny holes. The colony generally takes a U-shape with the mouth inside a crown of tentacles, which are lined on top, and the anus. As water circulates through the hollow tentacles from top to bottom, food is extracted from it. They live attached to the sea floor, or some submerged surface, with their tentacles freely floating in the water. Many types of bryozoa still exist today, mostly in tropical marine environments, although they are also found all over the world in freshwater, deep oceans and even in polar waters. There are around 4,000 known species, and the majority are colonial. Their exoskeletons are typically formed of calcium carbonate and so are usually well preserved in rock units. Their fossils can be seen in many areas in Oman, including Al Jabal Al Akhdar and Ras Al Had (Fig. 3.43).

3.3.11 Chitinozoa

Chitinozoans are marine microfossils with organic walls. These organisms are characteristically

Fig. 3.37 A collection of blastoids found by Alan Heward in Ras Al Had. Most of the samples are around 1–2 cm across. They represent a highly diverse group with various fragments of thecae, cups, basals and brachialis. It is believed that many of these specimens will reveal new species of blastoids that have never been identified before. Some fossil samples have also never been reported outside West Timor

Fig. 3.38 On the top, a living black sea urchin next to a Moray eel. The urchin has long spines and a spherical test and was found close to the Ad Dimaniyyat Islands in northern Oman. Sea urchins are very common along the coasts of Oman and can be seen in all the coral reefs. They move slowly and feed on algae and invertebrates. Below are different sea urchin fossils from the Cenozoic rocks of Oman. Generally, fossils of echinoids are common in all the Phanerozoic rocks of Oman. The samples are about 5 cm long

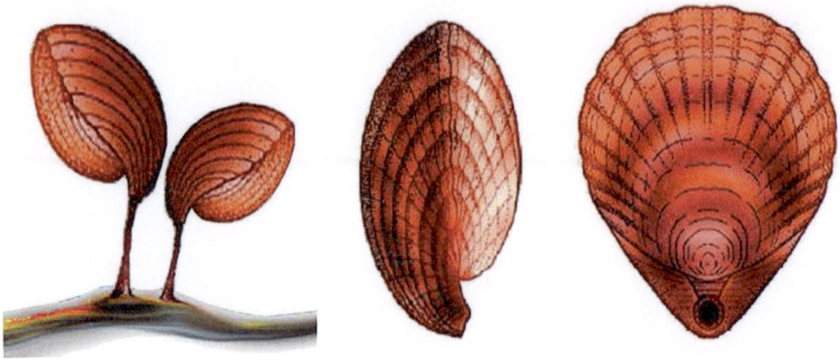

Fig. 3.39 Reconstruction of brachiopods, with a frontal view of the two valves and a view of the lower or brachial valve on the right. The brachiopods attach their stalk or pedicle to a substrate. The pedicle is attached to the upper valve (the pedicle valve). The pedicle valve is typically bigger than the lower brachial valve. The two valves are held together by a tooth and socket arrangement in articulate brachiopods and a system of muscles in inarticulate brachiopods

Fig. 3.40 Top left: a living brachiopod, picture from Cochise Collage. Other pictures show different types of Palaeozoic brachiopod fossils from the northern part of the Al Huqf area of Al Wusta Governorate and from Ras Al Had in the Wilayat of Sur. The biodiversity of the Ras Al Had samples is greater than that of Al Huqf. They are well preserved and represent different species of Permian brachiopods. Each brachiopod sample is about 5 cm long. The Permian fossils of brachiopods can record climate change in the Permian sea, from glaciation to subtropical

Fig. 3.41 On the top: trace fossils of tabular-like trees made by algae-eating animals found in the Cambrian rocks of the Wilayat of Muhut. In the middle: Skolithos of the Ordovician Period: vertical and cylindrical burrows, around 40 cm long and 1 cm wide. Bottom: Daedalus: garlic-clove-like trace fossils

similar to flasks, and their classification and origin remain unknown. They could be types of plants, eukaryotes or fungi. They could equally be eggs or represent the early stages of marine animals. They first appeared during the Early Ordovician and became extinct in the Late Devonian or Lower Carboniferous. During this period, their organic walls developed and varied significantly to form a variety of ornaments and shapes (Fig. 3.44). Many scientists suggested that chitinozoans could be linked to graptolite reproduction. This is primarily because the chitinozoans and graptolites lived during the same period in the Palaeozoic and their fossils normally occur in proximity and with the same high frequency. Microscopic chitinozoans are very abundant in the Ordovician and Silurian rocks of Oman, and they are often used to date and correlate subsurface rocks by petroleum geologists.

Four new chitinozoan species have been discovered recently by Al-Ghammari et al. (2010). Two of the new chitinozoan species are known as Belonechitina ghabahensis and Desmochitina omanensis, after Al Ghabah and Oman, respectively. They were described and illustrated from rich populations of Chitinozoans recovered from core samples at the Al Ghabah-1H1 petroleum well. Overall, at least one hundred species of Chitinozoa are found among the rocks of Oman.

3.3.12 Protista Kingdom

The Protista kingdom includes a diverse and very large group of microscopic or near-microscopic organisms. It is traditionally divided to two main types: the animal-like protozoa and the plant-like protophyta or algae. The protozoa are similar to animals in their high diversity and behaviour, such as movement; therefore, they are unconventionally known as primitive or first animals, which is the meaning of the name 'protozoa' (proto: first, and zoa: animals). Since most types of algae are today classified as plants, the Protista kingdom is often called as the Protozoa kingdom. The Protozoa could rather be described as a group for highly diverse unicellular eukaryotic micro-organisms that do not necessarily share a common evolution history. The single cell controls the processes, such as reproduction, feeding, excretion and growth. Among the most common types of protozoa are the foraminifera, radiolaria and dinoflagellates. The fossils of these organisms are found in Oman in the shallow and deep marine deposits that span the time period from the Early Palaeozoic to the Cenozoic.

Fig. 3.42 Trace fossils. Straight and inclined marks (yellow arrows) are present beside wave ripples in the Ordovician rock units in Al Amrat. These lines are probably formed by Ordovician organisms. They were probably something being blown across the sand flats on a thin film of water. Moreover, the rock bed also displays rounded shapes (black arrows) that might also represent trace fossils of extinct animals

3.3.12.1 Foraminifera

Foraminifera or forams represent a phylum of eukaryotic (single-celled) organisms that belong to the Protista kingdom. The name foraminifera means hole bearers. They range in size from about 20 μm to many centimetres in length. They are well known for their widespread, minute and simple-formed shells made of organic compounds, calcite crystals or cemented grains of sand. These shells often resemble tiny snails. Since their appearance, most likely in the Cambrian Period, at least 275,000 types of forams have lived on earth.

Foraminifera evolved significantly in the Palaeozoic and their diverse shells can be seen in shallow and very deep marine rocks, where they lived in the past and continue to live today (Fig. 3.45). They are divided into two main types: benthic and planktonic. The first type is more abundant and lives within or on top of the seabed, whereas the planktonic forams float in the water and sink to the bottom as they die. The living and dead shells of forams colour vast areas of sea floor—and water—as they thrive in certain environments. Their fossils, particularly when the organic matter is preserved in an anoxic environment where dissolved oxygen is lacking (e.g. in deep marine environments), produce source rocks for oil and gas. Furthermore, foram shells represent an important supply of hundreds of tons of calcium carbonates, which are used by marine animals to build shells and coral reefs in seas and oceans.

Forams are considered one of the best and key indicators of palaeoclimates and how many nutrients were available in the sea. For instance, the ratio of oxygen isotopes in their shells can give a good indication of how cold the water was when the shell was formed. Beyond that, foram shells are very often used as guides or index fossils that mark the relative age of sedimentary rocks. Forams which lived in a narrow time period can help to fairly precisely determine the age of such rocks.

Fossils of foraminifera are very common in the Phanerozoic rocks of Oman. In fact, some of the white or yellow Cenozoic limestones that can be seen around Muscat are almost entirely made of forams. Since ancient times, such stone has been used as a construction material—in both Oman and elsewhere, like in the Pyramids of Giza. The oldest widespread foram fossils in Oman are probably from the Late Palaeozoic, particularly the Permian Period. The rocks from

Fig. 3.43 Top left: living Bryozoa (freshwater species) in the Netherlands. Photograph by Rob Kool. Top right: a fossil of Permian Bryozoa in Al Jabal Al Akhdar in Nizwa, photographed by Fahad Al Thani. Bottom: a number of Bryozoa fossils from Ras Al Had in Sur. These fossils date back to the Permian Period. Generally, the fossil record of Bryozoa in the world goes back to the early Ordovician Period; therefore, they are the last major phylum of animals that appeared in the fossil record. They form colonies of individual members of clones, known as zooids, that are normally less than 1 mm long. The length of whole colony of zooids ranges from 1 cm to more than 1 m, and they vary significantly in shape as well

this period from—for example, Ash Sharqiyah South and Al Dakhiliyah—contain a variety of shells from a foram's type known as fusulinida, which died out during the Late Palaeozoic mass extinction. They are up to 0.5 cm long.

3.3.12.2 Radiolaria

The Radiolaria are a phylum of microscopic unicellular organisms that have external mineral skeletons and belong to the Protista kingdom. They are similar to foraminifera, but the radiolarians have skeletons composed of silica, compared to the calcareous (i.e. composed of calcium carbonate) external skeletons of foraminifera. Radiolaria live as planktonic organisms that float in the euphotic or sunlight zone of oceanic water where silicate or silicic acid forms an important nutrient, as opposed to carbonates that predominate in shallow sea water and make up the skeletons of foraminifera. The siliceous skeletal remains of radiolarians form a large proportion of deep oceanic sediments. As the organisms die, the skeletons sink to the ocean floor and cover it with a siliceous mat known as Radiolarian Ooze. Radiolarian fossils form 30% or more of the deep oceanic chert which is why deep oceanic sediments are often known as radiolarian chert, after the organisms that contributed to their formation. The Radiolaria first appeared during the Cambrian Period in the Early Palaeozoic, and they still live in massive quantities in ocean waters, although 90% of radiolarian types are extinct.

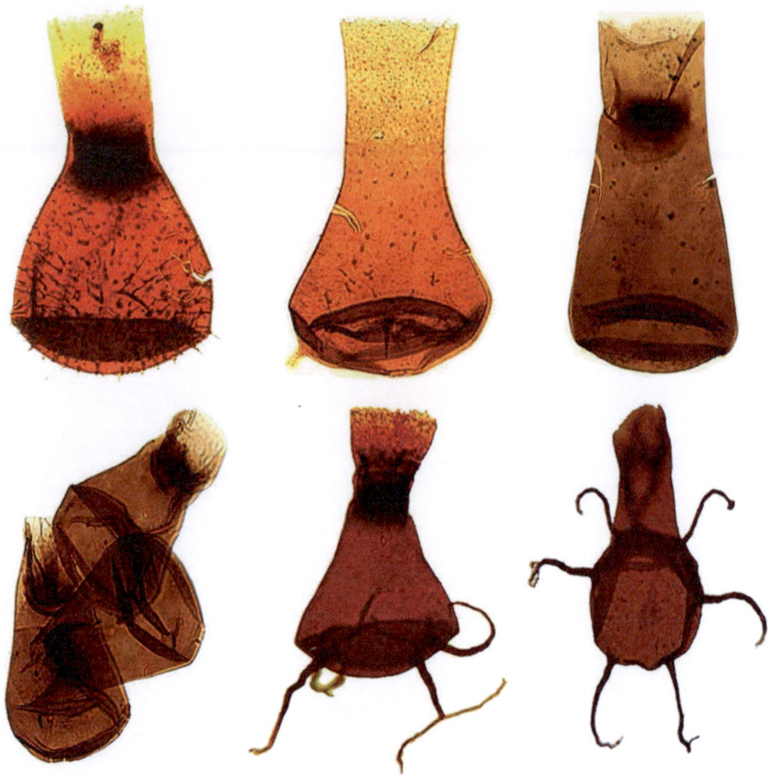

Fig. 3.44 Different microscopic forms of Chitinozoa, the flask-shaped organic-walled marine microfossils. These samples were found by the geologists Mutasim Al-Ghammari and Graham Booth from Petroleum Development Oman. The fossils belong to the Early Palaeozoic. Their wide distribution in the marine rocks of Early to Mid-Palaeozoic makes them very useful to correlate between the layers of rocks

However, since their appearance in the Cambrian Period, Radiolaria have rapidly evolved and have produced many diverse species, particularly during the Jurassic and Cretaceous periods of the Mesozoic Era. Their fossils are often used as index or diagnostic fossils to define the age of oceanic sedimentary rocks and to understand changes in ancient climates. Radiolarian skeletons are often attached with a group of needle-like false feet or pseudopods that give the organism buoyancy in the water and function to catch food particles. Radiolarians may also contain symbiotic algae that use photosynthesis to provide energy for the radiolarian cells.

Radiolaria lived in the waters of the Neo-tethys Ocean. Oman and the Arabian Plate were located on the western side of this ocean (see, e.g. Fig. 3.46). Huge amounts of radiolarian skeleton remain littered the floor of this ocean to form the primary deposits of what are known as Hawasina rocks or the deep oceanic sediments of the Neo-tethys Ocean. The fossils or Radiolaria in these rocks belong to the time between the Permian Period of the Late Palaeozoic Era to the Cretaceous Period of the Late Mesozoic. They have significantly contributed to the dating of the deep oceanic chert deposits that normally lack other types of marine fossils.

3.3.13 Conodonts

Conodonts represent a class of extinct animals that belong to the phylum Chordata. The name is derived from the Greek words konos for cone and odont for tooth. Their fossils are usually remains of small teeth-like fragments. The fossil fragments represent the feeding apparatus of the actual animals. These animals are believed to be eel-like creatures that appeared during the Palaeozoic or possibly the end of the Precambrian and became extinct at the mass extinction between the Triassic and Jurassic periods about 200 million years ago, during a period when

Fig. 3.45 Different fossils of foraminifera as they appear to the naked eye and under the microscope. They belong to the Palaeozoic, Mesozoic and Cenozoic Eras of the Phanerozoic

dinoflagellates and many other nannofossils became widespread (Fig. 3.47). The fossils of conodonts remained an enigma for a very long time, despite their importance in determining the ages of Palaeozoic rocks. However, this mystery was resolved when rare whole-animal fossils of conodonts were discovered alongside the teeth-like ones. Nevertheless, actual information about these animals remains sparse and their classification as chordates or vertebrates remains in dispute. There are about 1,500 different species of conodonts discovered so far. However, they have been identified through the morphology of the microfossils of conodonts so represent an unusual classification. The primary and common fossils of conodonts have a simple cone that was possibly formed by a single tooth. Other shapes of fossils could be blade-like, bar-type or a platform. The average size of these fossils ranges between 0.2 mm and about 5 mm, so they are very rarely seen by the naked eye. They are usually extracted from rocks using acids which dissolve the host rock but not the conodont fossils. Conodonts appear to contain large eyes and flaps. The lateral position of the eyes in the head suggests that conodonts were not predators. The total length of the animal might reach many centimetres.

Fossils of conodonts can be found in the Palaeozoic rocks of Oman, particularly in the rocks of the Ordovician Period (Fig. 3.48), which is the period when conodonts became diverse and widespread. The Ordovician rocks

Fig. 3.46 Various fossils of Radiolaria from Oman. They belong to the Palaeozoic and Mesozoic (De Wever et al. 225). Radiolaria are tiny planktonic organisms that drift in oceans and seas and feed on bacteria, zooplankton (water animals) and phytoplankton (sea plants)

could be sited in Wadi Daiqa and in the wadis of Al Amrat. Currently, a number of researchers are working to extract and study conodont fossils from Oman to correlate them to subsurface data and to understand more about these interesting creatures. Conodont fossils are also found in Triassic rocks in different parts of Oman. The scope of study is wide open to know more about conodonts through the information stored in the rocks of Oman Figs. 3.49 and 3.50).

3.3.14 Ostracods

Ostracods represent a class of crustaceans animals that belong to the arthropod phylum, similar to lobsters and crabs. They live as benthic animals or less commonly plankton in sea water or freshwater. Some types of ostracods also live in wet forest soil.

The name is derived from the Greek word ostrakon, meaning shell. These animals are

Fig. 3.47 Reconstruction of a toothy conodont by Bill Parson

Fig. 3.48 Fossil-rich beds with corals, brachiopods, crinoids and bivalves from the Ordovician Period. Recently, fossils of Ordovician conodonts have been extracted from these rocks

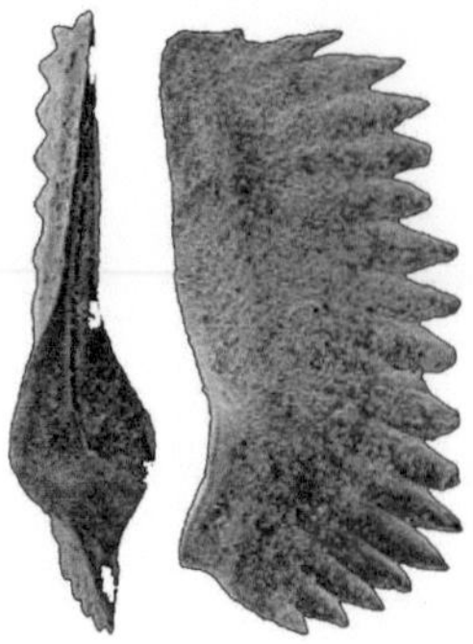

Fig. 3.49 Various fossils of conodonts from the Triassic Period of Oman (Orchard 1995). These samples are well preserved, and they include new species of Lower Triassic conodonts that led to a taxonomic revision of previous classifications

Fig. 3.50 Fossils of conodonts from the Permian Period. They were recovered from the Wilayat of Ar Rustaq and Al Mudaybi. Some are Mesogondolella omanensis, representing new species from Oman (Kozur1 and Wardlaw 2010)

considered among the most common and most fossilized arthropods since their appearance in the fossil record in the Cambrian and Ordovician periods of the Early Palaeozoic up to date. More than 65,000 species of ostracods have been identified, 13,000 of them still alive today. Their lengths usually range from 0.2 mm to 3 cm. Because of the good preservation of ostracods fossils in comparison with other crustaceans and they are also among the most prevalent arthropods that lived in the Palaeozoic, ostracod fossils are considered very useful in understanding the changes in palaeoclimates and for the local and regional correlation and biozonation of Palaeozoic marine rocks.

Bodies of ostracods are composed of a head and thorax (chest) (Fig. 3.51). These two parts are separated by a small constriction. The head normally represents the largest part of the body, and it is attached with a pair of mandibles, two pairs of maxillae and two pairs of antennae that help the animal to swim in water. The body is encased within two hard valves. The appendages of the soft internal body emerge from the hard valves. Once the animal is faced with danger, the two valves firmly close using an internal muscle system. Species of ostracods have a wide range of diets. They could be herbivores, carnivores, filter feeders and scavengers. In turn, there are also many animals that eat ostracods.

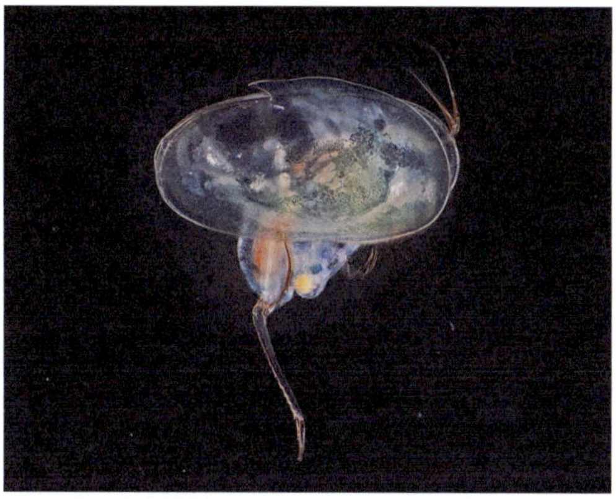

Fig. 3.51 A living ostracod, photograph by George Hetcht. The animal is few millimetres long. These animals appeared in the Early Palaeozoic and continue to dominate the seas today. They are moderately diverse in the seas of Oman today, with about 100 different species of ostracods reported. Ostracods have different techniques to survive in harsh conditions, even when taken out of water or when they get eaten by fish. The eggs of ostracods can also survive for many years when water completely dries up

Fossils of ostracods are present in all time eras of the Phanerozoic Eon. New species are continuously discovered in the northern and eastern seas of Oman. The fossils of these ostracods have provided a lot of information about ancient climates around Gondwanaland and the Palaeo-tethys and Neo-tethys oceans in the Late Palaeozoic, aiding our understanding of the micro-continents that affected Gondwana and how the continents coalesced (Fig. 3.52).

3.3.15 Copepods

Copepods belong to the crustacean subphylum, like ostracods. They represent a large group of prevalent small arthropods that live as planktonic or benthic animals in oceans, shallow seas and almost all freshwater habitats, both warm and cold. Some types also live in wet terrestrial habitats, such as springs, swamps, forests and caves. Although copepods have been widespread for millions of years, they fossilize poorly, primarily because of their small size which rarely exceeds 1 mm and because of the fragility of their bodies which tend to disappear as time passes (Fig. 3.53). Previously, scientists thought that the oldest fossil records belonged to the Cretaceous Period of the Mesozoic; however, recent fossil findings suggest the appearance of copepods in the Early Palaeozoic and possibly the end of the Proterozoic.

There are at least 13,000 identified species of copepods. About 2,800 of these species live in freshwater. In many cases, identifying copepod species is a not an easy task as it requires a lot of detailed analysis.

The bodies of copepods consist of multiple thoracic segments and a relatively big head that fuses with the top thoracic segments. The head and thorax are often transparent and encased within a resistant exoskeleton. Two pairs of antennae emerge from the exoskeleton which

Fig. 3.52 Ostracods from the Permian Period from Crasquin-Soleau, 2003. They have been recovered from the Permian rocks of Al Huqf area in Central Oman. Ostracod fossils from the Permian rocks of Oman and elsewhere (e.g. South China) around the Tethys Ocean have supported new interpretation of the palaeogeography and the reconstruction of the ancient Pangaean supercontinent, which formed during that period. They also indicate that Oman was located in tropical latitudes at that time. Samples are magnified more than 50 times

helps the animal to move and feed. Copepods also possess one or two eyes in the middle of the head and have an abdomen and excretory system. They feed on bacterial deposits, algae, plankton or other smaller copepods. In turn, they get eaten by fish and many other marine animals. Although copepods respond very fast to attacks, predators often approach them from behind, swimming slowly to avoid inducing turbulence that would alert the copepods. Through history, copepods have played an essential role as a food source for many animals and in the carbon cycle, absorbing millions of tons of carbon emissions every year.

Researchers have recently found copepod fossils from Oman that most likely date back about 303 million years. Interestingly, they were recovered from a solitary bitumen rock fragment within a glacial deposit of the Late Palaeozoic in the Al Huqf area of Central Oman (Fig. 3.54). This is probably the first time that soft-tissue fossils have been recovered from isolated bitumen clasts. The fragments are well preserved and represent parts of the limbs and antenna of copepods. They also include fossils of intrinsic muscles that show strands. They have pushed back the global fossil record from previously

Fig. 3.53 A living copepod, photo by George Hetcht. The animal is normally 1–2 mm long. There are at least 30 different species of copepods living offshore around Muscat in shallow seas of around 10 m. These animals play an important role in the ecosystems of Oman's coastal areas as they can be major consumers of harmful algal blooms and they also part of the food chain as prey for larger marine animals. In general, most copepods have a single median compound eye (red in colour as shown in the figure) and two pairs of antennae. The eggs of many copepod species are enclosed in a sac and attached to the female's body until they hatch

Fig. 3.54 A bitumen clast in the Al Huqf area in Central Oman. The clast is believed to be around 303 million years old, and it contains a number of the oldest known copepod fossils in the world (photograph after Alan Heward)

discovered Cretaceous fossils by some 188 million years, and of free-living forms by 289 million years. They probably lived in a glacial lake where oil (the bitumen clasts completely lack biodegradation) leaked into the lake from underlying rocks. It is worth noting that the glacial deposits of the Late Palaeozoic in the Al Huqf area are close to the oil field in southern Oman, and glaciers most likely moved from South to North Oman during this era (Fig. 3.55).

3.3.16 Plants

Plants first appeared on land as algae during the Early Palaeozoic. Grains of the spores and

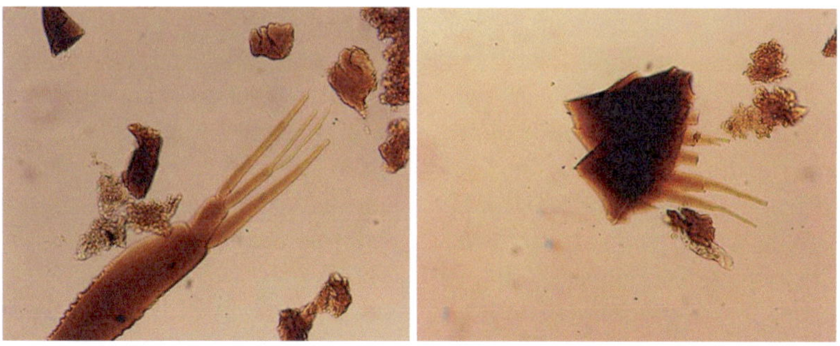

Fig. 3.55 Crustacean fragments from upper Carboniferous bitumen, most likely of copepods (Selden et al. 2010). These are probably the oldest known fossils of copepods. Previously, the oldest known fossils were from the Cretaceous Period (about 115 million years old), whereas these are from the late Carboniferous (about 303 million years old). Their discovery in Oman has extended the fossil record of copepods by about 188 million years. In general, fossil copepods are uncommon. Although these crustacean animals are very abundant, they poorly fossilize

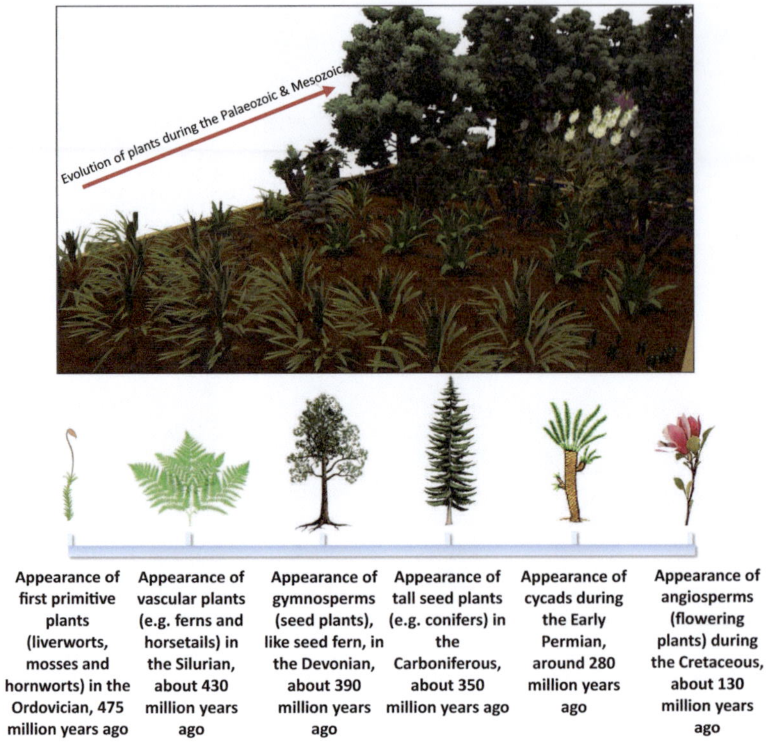

Fig. 3.56 Evolution of plants during the Palaeozoic and Mesozoic. Nearly all stages of evolution can be witnessed in the rocks of Oman. The first plants on land are most likely algal mats and liverworts followed by the more complex gymnosperms and angiosperms of the Mesozoic which later became the most dominant plants on land

pollens of these early plants are preserved in the surface and subsurface rocks of Oman. In fact, the rocks of Oman preserve almost all the chapters of plant evolution, from their first emergence to the evolution of angiosperms (Fig. 3.56).

The story of plants begins in the Ordovician Period as liverworts and mosses which reproduced through spores. This was followed in the Silurian Period by the emergence of seedless vesicular plants with their beautiful leaves, such as ferns. Shortly after, in the Devonian, gymnosperms appeared. Their name comes from the Greek words gymnos and sperma, meaning 'naked' and 'seed'. Examples include Glossopteris, a large tree fern, now extinct, which grew mainly in tropical environments. By the Carboniferous time, large trees, like conifers, had formed large, dense forests. At the same time, cycads, which resemble pineapples or palm trees, also appeared. Later, in the Permian Period, another type of gymnosperm, known as ginkgo, also evolved. The fossil seeds of these trees can be found in the Permian rocks of Oman. The final chapter in plant evolution was the emergence of angiosperms, including flowering plants, during the Mesozoic Era. They take their name from the Greek words for 'receptacle'—angeion—and 'seed'—sperma—as their seeds are enclosed.

The rocks of Oman have preserved—in great detail—much of the evolutionary progress of plant life. They are an important scientific resource in understanding the climatic, tectonic and biological changes that occurred on earth and were captured in the extensive fossil record. This record includes spores, pollen, roots, leaves and even large tree trunks. Apparently, in its long history, Oman was appropriately located at a latitude which engendered the growth of such prehistoric giants.

3.3.16.1 Primitive Plant Spores (Cryptospores)

The earliest known record for the appearance of early land plants goes back 475 million years to the Ordovician Period. It was recovered from the subsurface rocks of a petroleum well, Al Ghabah-1, in Central Oman (Fig. 3.57). It represents fossilized

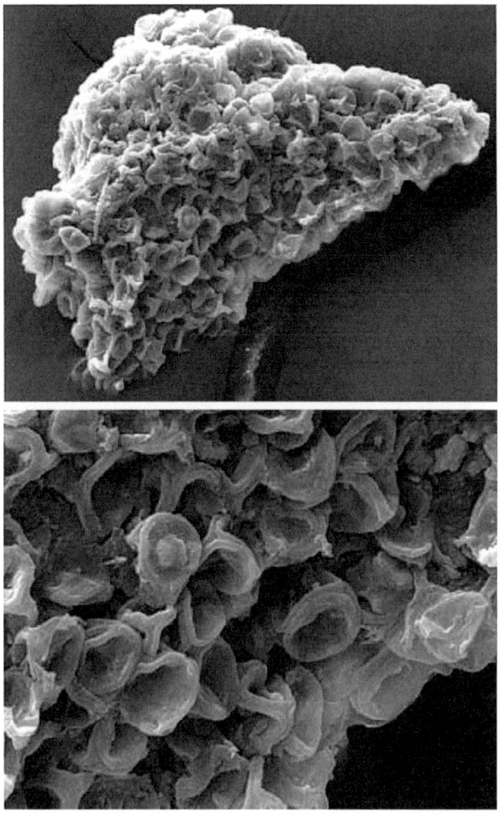

Fig. 3.57 Fossil cryptospores. On top: tetrad cryptospores found in rock samples extracted from Al Ghabah-1 well by Uzma Mohiuddin from Petroleum Development Oman. The image appeared in Nature (Wellman et al. 2003). Below is an enlarged image of the above image. The age of these samples is estimated at 475 million years, of mid-Ordovician age, a time in which Oman was located on the north-west margin of the supercontinent Gondwana. These microscopic dispersed spores represent the oldest known fossil evidence of plants in the world. They are derived from terrestrial plants. The analysis of spore wall ultrastructure supports liverwort affinities

microbial and primitive plant spores known as cryptospores, from the Greek words cryptos, meaning 'hidden', and sporum, meaning 'seed'. Cryptospores from Oman represent the oldest known evidence of the migration of plants from water to land. They show concentric lamellae or thin sheets in their walls which suggests that they originated from liverworts of no more than a few centimetres in height. These primitive plants appeared on land following a change in the concentration of atmospheric gases similar to their percentages today. Oxygen increased and toxic gases significantly reduced compared to the Precambrian. In addition, the earth's climate became wetter and more capable of sustaining plant life.

Cryptospores exist in many different forms in the Ordovician and Silurian rocks of Oman (Fig. 3.58). They may largely be classified to three main shapes: monads with one spore, dyads with two spores and tetrads with four spores. Dyads and tetrads may be enclosed within outer envelopes with different contacts between the spores. They probably disappeared in the Silurian Period, whereas monads continued for some time. It is likely that the spores had walls made of many lamellae which were often dispersed across vast distances. Although they are a product of land plants, they have also been found in marine deposits because of their dispersal by wind and water.

3.3.16.2 Spores

The first evidences of plants on land are Ordovician monad, dyad and tetrad cryptospores of liverwort in the mid-Ordovician, about 475 million years ago. This was followed by the appearance of spores from early vascular plants in the Upper Ordovician and Silurian. As time passed, larger and more diversified spores evolved. Together with pollens, plant spores are the basis of the main method of dating and correlating the Carboniferous and Permian rocks of Oman because of their rapid evolution. Spores of ferns and mosses are among the most common ones during that period. These were dispersed in vast numbers from a spore capsule by water and wind. Hence, plants were able to grow in large numbers and dominate large areas of land during the Palaeozoic. Because of this wide dispersal, spore fossils can also be recovered from marine deposits. The record of Palaeozoic spores from Oman also marks continuous reworking of deposits by repeated glaciation and meltout conditions when Oman was close to the South Pole in the Palaeozoic. With younger times in the Palaeozoic, this record displays a warming and drying trend from the Carboniferous to the

Fig. 3.58 Different microscopic cryptospore fossils identified in the laboratories of Petroleum Development Oman. From right to left: Monads, Dyads and Tetrads. Cryptospores represent the oldest evidence of plants on land as they appeared in the mid-Ordovician, some 475 million years ago. They resemble microstructures of living liverwort spores today and they were produced singly (monads), in pairs (dyads) or in group of four (tetrads)

Permian periods, as Oman raced northward to warmer climes (Fig. 3.59).

3.3.16.3 Pollens

During the Carboniferous Period, the gymnosperms or naked-seed fern plants were very prevalent. Pollens of different types were, therefore, widespread in the rocks deposited during that period and after. They probably appeared in the Devonian, but kept increasing in abundance with time. The most common types of pollen of the Late Palaeozoic are the monosaccate pollens. Their spread was supported by frequent rainfall so that different types of ferns, such as glossopteris, and trees, such as conifers, grew in swamps and wetlands. Later, they also became common in drier areas of land before they became extinct at the Late Palaeozoic mass extinction. Fossils of pollens are used as index fossils to correlate different rocks and determine their ages. During the Palaeozoic glacial periods that affected Oman when it was located close to the South Pole, many organisms could not survive the cold, so pollen fossils are possibly the only way to know the age of rocks. They also give a good indication of the climatic conditions during their depositions. Their exquisite forms are clear under a microscope lens (Fig. 3.60).

It is interesting to note that spores and pollen found in rocks that were deposited in glacial conditions might not represent plants that were living at the time when the glacial rocks were deposited, but most probably the vegetation that

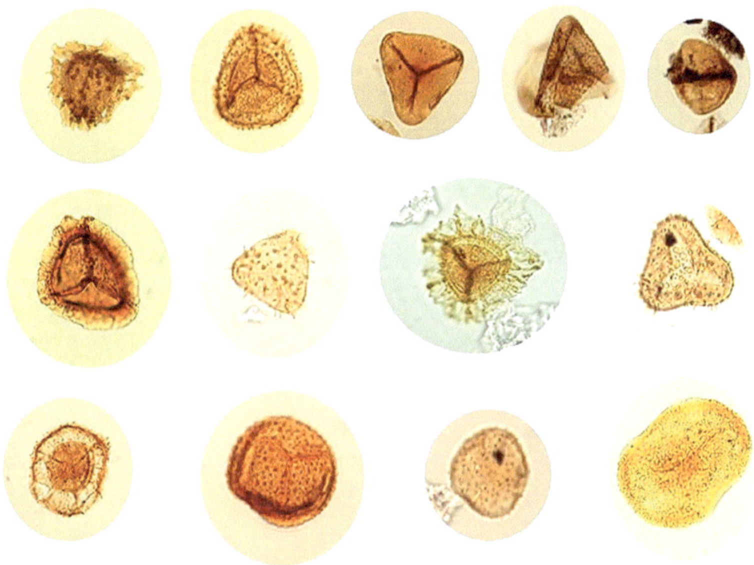

Fig. 3.59 Representative spores from the Carboniferous rocks of Oman (Penny and Osterloff 2002). The record of Carboniferous spores in Oman is a reflection of changing condition between glacial maxima when Oman was quite close to the South Pole. Samples are about 100 μm across

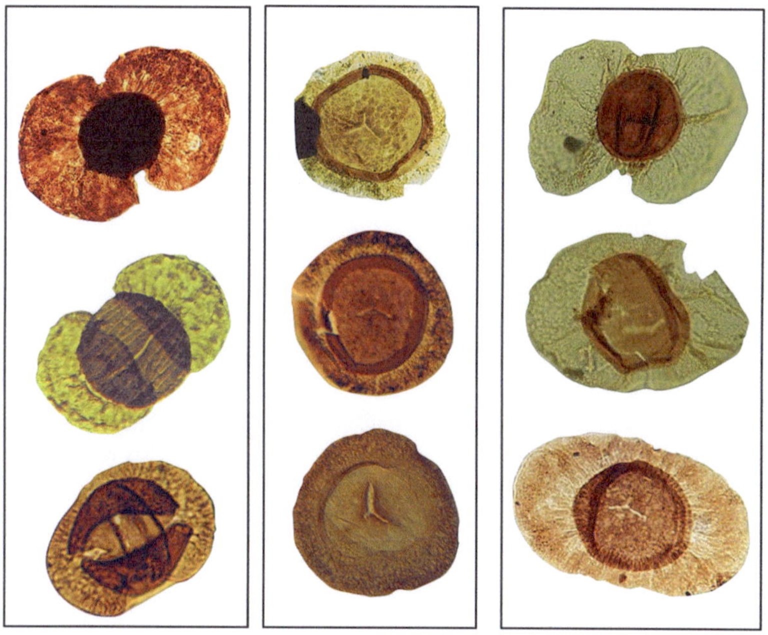

Fig. 3.60 Palaeozoic gymnosperm fossil pollens identified by the geologists of Petroleum Development Oman. On right and middle: Monosaccate pollens of one bladder or sac with lateral symmetry and radial symmetry. On left: Bisaccate pollens that have two bladders. The last pollen is usually produced by conifer trees

was present during a previous interglacial period. Spores and pollen are remarkably tough and can be readily reworked.

3.3.16.4 Trees

In several locations in the Governorate of Al Wusta, there are large groups of petrified trees that date back to the Permian Period, and are as much as 270 million years old (Fig. 3.61). Collectively, these sites are considered one of the largest concentrations of petrified trees from the Permian Period in the world, and certainly the largest in the Arabian Peninsula. They grew on river banks and along streams that most likely

Fig. 3.61 Permian tree trunk fossils from the Governorate of Al Wusta. The above petrified wood is clearly in situ and it is buried in tight grey sediments. The process of wood petrification starts with the burial of logs in tight sediments that lack oxygen (an anaerobic environment), which inhibits the process of decomposition. This allows the slow replacement of organic matter (cells of lignin and cellulose) by hard minerals like silica which is carried by mineral-rich water that flows in the sediments. Eventually, the organic matter is completely replaced by minerals, but its structures are supported by hard replacing minerals. The fossil trunks in these pictures are not compacted and so the process petrification have likely occurred during shallow burial

flowed from the south-eastern part of Oman towards the north-west. Some of the petrified tree trunks in the wilayats of Muhut and Haima are at least 15 m long. These trees apparently fell along the direction of flow of the water and are often buried in perfect alignment. Since the lower part of the fossilized tree can be distinguished from the upper part because of its width, it is possible to determine the direction in which the trees fell, and therefore the direction of flow of the watercourse concerned—in this case, southeast to northeast. The soil in which some examples are embedded contributed to their amazing state of preservation. It allowed the replacement of the wood matter—over great lengths of time—by the hard and weather-resistant silicate minerals. This captures the structure of the tree in minute detail.

Details are important in understanding life on earth during the Permian (Fig. 3.62). For instance, the number of rings on in a tree trunk can indicate the age of a tree when it died. Wide rings are a reflection of a good growing season, while narrow ones often indicate a dry year, or poor conditions. The cross sections of the fossil trees also reveal radial rays, known as vascular rays, extending from the pulp of the trunk towards the bark and divide the stem to pie-like slices. These rays are responsible for the radial transport of dissolved nutrients and water across the stem. In some trees, the tissues of the bark—the outermost layer of the trunk—are also well preserved. The fossilization process probably took tens or hundreds of years. During this period and in the absence or lack of oxygen, the

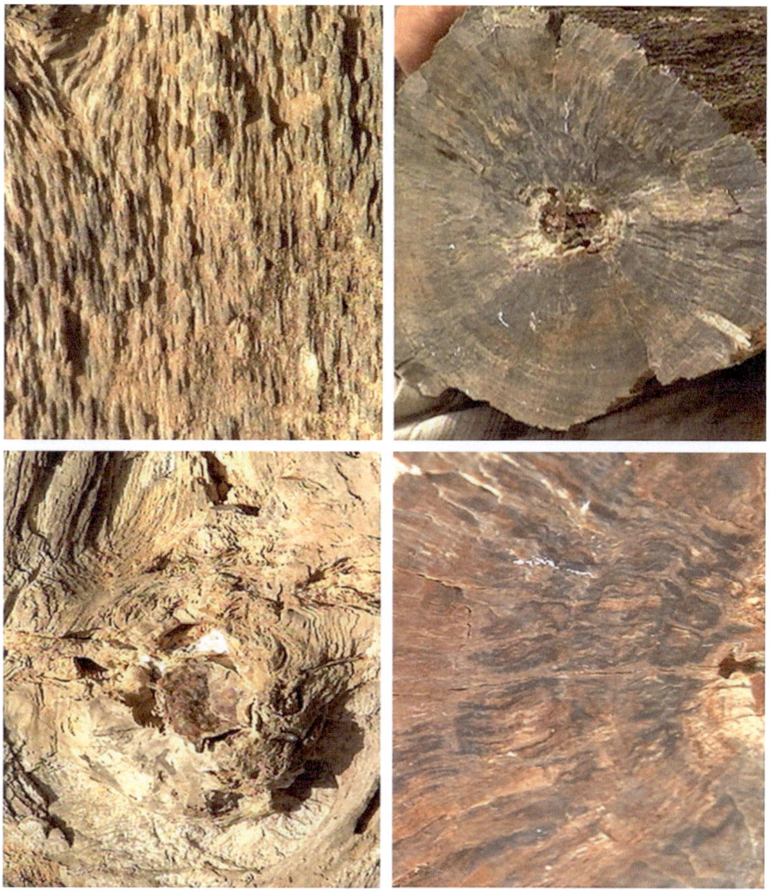

Fig. 3.62 Close-up photographs at various parts of a petrified tree trunk that shows the growth rings and pulp (top right), the outer bark (top left), the radial rays extending to the bark (bottom right) and a knot (bottom left), which indicates a place in the trunk where a branch grew. Petrified wood can preserve the original structures of trees down to the microscopic scale, and they can provide a lot of information about the Palaeoclimate. Insect burrowing can be seen in some of them

organic matter remains for a long time without decomposing. This allows the silicate minerals—particularly quartz—which are usually resistant to weathering, to replace the organic matter and preserve the original organic texture of the tree. In many cases, fossil fruits of the Permian trees were also found (Fig. 3.63).

3.3.16.5 Roots

The Permian sandstone rocks of Al Wusta Governorate contain what appear to be the fossil roots of Permian trees. They are sited at the edges of cliffs in the wilayats of Ad Duqm and Muhut. They most likely represent ancient soil that formed in the humid, tropical climate of the Permian Period, in which giant trees grew. This ancient or palaeo-soil (often known as palaeosol) forms successive layers of beautifully coloured rocks that reflect the varied environment of wetting and drying during the Permian (Fig. 3.64).

3.3.16.6 Leaves

The Permian rocks of the Wilayat of Muhut consist of a layer of grey stone about one metre thick. This layer contains a lot of fossil leaves and twigs that were deposited some 270 million years ago. The Permian plants probably lived in extensive swamps before they died and turned into organic matter that was preserved for long time because of the absence or lack of oxygen in the soil (Fig. 3.65). They were most likely ferns and conifers, which dominated the Gondwana

Fig. 3.63 Pine fruit fossils (pine cones) found near the sites of petrified trees in Al Wusta Governorate. The fossil is shown alongside similar present-day pine cones. In the middle, a compacted example lacking detail, although the top of the cone is still visible. The top one is found by Jan Schreurs, the middle one is found by Al Khattab Al Mahruqi and the lower one is found by Andrea Marasciello. Conifers became very dominant during the Permian Period and they spread inland and on mountains, forming large conifer forests. The fossil pine fruits come in different shapes, some have long and slender shape, others are egg-shaped and few are thick and round. The sample is generally around 10 cm long

supercontinent in the Late Palaeozoic, together with Glossopteris, a fern-like tree which could reach thirty metres in height, but which died out during the extinction event at the end of the Palaeozoic.

3.3.17 Summary of Life in Oman During the Palaeozoic

Line drawing with the Palaeozoic fauna and flora found in Oman is shown in Fig. 3.66. From top to bottom: a generic trilobite, sacabambaspis fish,

Fig. 3.64 Possible fossil roots of Permian plants (where the black arrows indicate) that developed in ancient soil deposited at the end of the Palaeozoic Era

generic sea urchin, alatoconchid bivalve, solitary and colonial coral, orthocone and coiled nautiloids, generic brachiopod, generic crinoid and generic blastoid. The plants represented are (from top to bottom) as follows: a primitive hornworth, horsetail and fern, seed fern, tall conifer tree and a cycad.

The alternation of marine and terrestrial deposits in Oman during the Palaeozoic has preserved both the sea and land organisms of the period. Oman at that time was initially part of the Gondwanaland and later part of Pangea. It witnessed many tectonic events and was subject to different climates while being at a strategic position on the edge of ancient supercontinents, allowing this land to record the environmental history and evolution of life during the Palaeozoic.

Appearing during the Cambrian explosion of life and becoming extinct during the Permian mass extinction, trilobites existed throughout the

Fig. 3.65 A varied set of fossilized leaves from the Permian (about 270 million years old) in the Wilayat of Muhut. They form a mixture of various types of leaves from the glossopteris trees (an extinct order of seed ferns), conifers and tingia floras that lived in the Gondwanaland during the Permian Period. The leaves of glossopteris are tongue-shaped (like the bottom left fossil); hence, the name is derived from glossa which means 'tongue' in Greek. The samples are 5–10 cm long

whole Palaeozoic Era and represent the most prominent organisms of the time. Fossils of trilobites have been found in different places across Oman. Their fossil record in Oman goes back more than 500 million years to the Cambrian Period and extends to the Permian Period—a period of almost 300 million years. New species are always being discovered.

Recently discovered in Oman, primitive fish fossils of an Ordovician species known as Sacabambaspis are just one of a range of specimens reflecting the scope and diversity of the Palaeozoic oceans. Organisms included crinoids, nautilus, bivalves, brachiopods, blastoids, corals, conodonts and graptolites. Fossils of foraminifera, radiolarian and dinoflagellates that belong to the Protista kingdom are also very common in the Palaeozoic rocks of Oman.

The rocks of Oman have preserved—in great detail—much of the evolutionary progress of plant life. They are an important scientific resource in understanding the climatic, tectonic and biological changes that occurred on earth and were captured in the extensive fossil record. The earliest known record for the appearance of early land plants on earth goes back 475 million years to the Ordovician Period, and it was recovered from the subsurface rocks of a

Fig. 3.66 Summary of life in the Palaeozoic Era as recorded by the fossil evidence from Oman, showing the evolution of animals and plants throughout the Palaeozoic. Trilobites, primitive fish of Sacabambaspis, blastoids and Glossopteris appeared and died out during the Palaeozoic. They were important parts of the ecosystems in water and on land and they contributed to the evolution of more diverse and complex life forms on earth. Overall, the Palaeozoic started with biggest radiation of life forms and ended up with the largest mass extinction of life forms the earth has ever witnessed

petroleum well in Central Oman. The record of plant evolution in Oman includes a wide variety of spores, pollen, roots, leaves and even large tree trunks of different plant types from the Palaeozoic. Apparently, in its long history, Oman was appropriately located at a latitude which engendered the growth of such prehistoric giants.

4 The Mesozoic Rocks and Life in Oman

The appearance of dinosaurs and their signficant widespread across the globe represents the most prominent life phenomena during the Mesozoic. This dinosaurs vertebra was found in the Al Khawd area in the Governorate of Muscat. It dates back to the Cretaceous Period — about 70 million years ago. It is about 15 cm long and was found by the author during a survey of the Geological Society of Oman (GSO) in 2014.

M. H. Al Kindi, *Evolution of Land and Life in Oman: An 800 Million Year Story*,
https://doi.org/10.1007/978-3-319-60152-6_4

4.1 Earth During the Mesozoic

The Mesozoic Era spans a time of 186 million years, from 252 million years ago to 66 million years ago. It is the second age of the Phanerozoic Eon. Because of the dominance of reptiles on earth, the Mesozoic is often known as the age of reptiles. However, life at the beginning of the Mesozoic was struggling to survive, following the worst ever mass extinction in the history of earth. It occurred at the Palaeozoic–Mesozoic boundary and is called the Permian–Triassic extinction or in short form P-T extinction or the Great Dying.

Nevertheless, the Mesozoic witnessed the emergence and spread of dinosaurs and the appearance of birds, mammals and flowering plants. The era ended with another mass extinction which killed off all the dinosaurs and many marine fauna, such as ammonites. It was a time of major climatic changes and dramatic tectonic events, both contributing to the evolution of life on earth. In fact, the environment during the Mesozoic was more suitable for evolution and diversity than at any time in the past.

The Mesozoic is divided into three periods of time, known as the Triassic, the Jurassic and the Cretaceous (Fig. 4.1). The first is named after three distinct rock units (tri means three) found in Germany and northwest Europe, and it spans the period from 252 million years ago to 201 million years ago. The Jurassic is named after the Jura Mountains in the Alps, between France and Switzerland, and lasted 55 million years from 201 million years ago to 145 million years ago. Finally, the Cretaceous Period spans a time of 80 million years—the longest time period of the Mesozoic (Fig. 4.1). The name is derived from the Latin creta, meaning 'chalk', as it is a period during which many of the world's chalk deposits were laid down. It is often abbreviated to 'K' after the German word for chalk, kreide.

The Mesozoic Era began with a dry climate and low sea levels worldwide. Initially, life struggled to regain a foothold after the P-T extinction. It took around twenty million years at the beginning of the Mesozoic for life to flourish

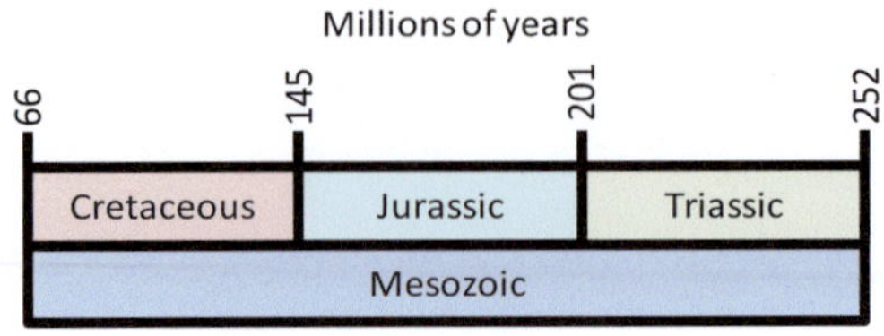

Fig. 4.1 Time periods of the Mesozoic Era

again. Although the P-T extinction had devastating effects, leading to the extinction of most animal species on land and sea, it did allow for the radiation of many new forms of life during the Mesozoic. Like a tree sprouting new branches, new species of reptiles, amphibians and molluscs became widespread. Dinosaurs and small mammals appeared among the cycads and ferns which dominated at the beginning of the era. Right across the globe, however, all forms of life were to experience some of the most turbulent times in earth's geological history.

During the Triassic, almost all the continental plates coalesced into the largest ever supercontinent in the history of earth, known as Pangaea. Pangaea had a ring shape, but was not completely closed on one side (see, e.g. Fig. 3.9). The inner ocean was known as the Palaeo-tethys Ocean, whereas the outer was named the Panthalassa from the Greek pan meaning 'all' and thalassa meaning 'sea'. It is also sometimes called the Palaeo-Pacific. Vast inland areas of Pangaea were far from the sea. The temperatures in these often-dry areas fluctuated between burning hot summers and freezing cold winters, resulting in seasonal temperature extremes. The Panthalassa and Palaeo-tethys, however, experienced less severe temperature differences, primarily because of the high-specific heat capacity of water. This temperature differential between the land mass and the oceans resulted in enormous turbulence during the Early Mesozoic, with gale force winds and massive storms. At the end of the Triassic, life had to suffer yet another mass extinction. It probably wiped out around 20% of all life species, particularly in the sea. This extinction, however, allowed many other species, like the dinosaurs, to diversify during the Jurassic and Cretaceous.

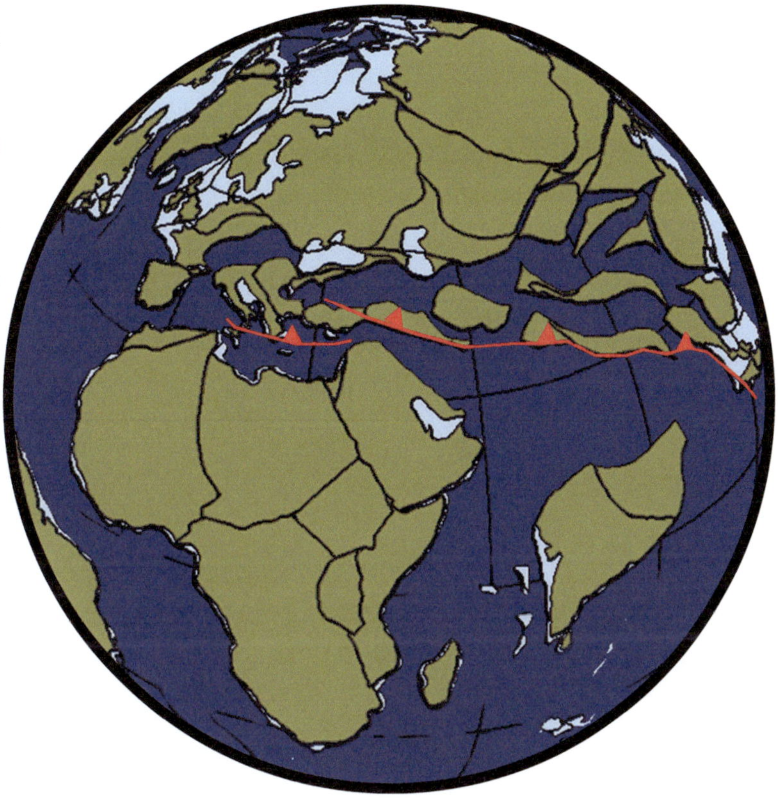

Fig. 4.2 Positions of continents across the globe at the end of the Mesozoic. Oman and Arabia were positioned near the equator. The Neo-tethys Ocean started to close at the end of the Mesozoic mainly due to the racing of Africa and Arabia to the north as a result of the opening of the Atlantic Ocean and the separation between South America and Africa

During the Jurassic Period, the Pangaean supercontinent started to break up into smaller continents, increasing the areas of shallow sea and coastal zones in the world (Fig. 4.2). The climate on land became less arid and more humid as their proximity to the sea increased. Deserts retreated and were often replaced by lush forests inhabited by the most thrilling animal of the time —the dinosaurs. The first birds also evolved during the Jurassic, with the appearance of Archaeopteryx which takes its name from the Greek words archaeos meaning 'ancient' and pterix meaning 'wing'. Butterflies, ants and other insects became common on land. Also, new species of reptiles, fish and amphibians emerged in the seas. Molluscs became enormously diverse as well, with the appearance of animals like rudists and belemnites and the blooming of ammonites. Great varieties of gastropods and corals thrived in the shallow seas. Then, amidst this burgeoning of life, came another setback. The Jurassic Period ended with a minor extinction phase. A few species of dinosaurs were wiped out as well as some marine species. The exact cause of this extinction is unknown.

As the last time period in the Mesozoic commenced, the continents were more or less distributed across the globe in the form we know today. The break-up of Pangaea continued and new oceans developed, such as the Atlantic Ocean between South America and Africa. The sea level rose, and global temperatures increased, covering vast areas of the continents with warm, shallow seas. It is estimated that a third of today's land masses were submerged by sea water. It was one of the best times ever for marine organisms to flourish and diversify. Marine animals, like rudists and ammonites, thrived and dinosaurs continued to roam the remaining exposed dry lands of the continents. The separation and drifting apart of continents during the break-up of Pangaea promoted the evolution of different organisms in different parts of the globe. New species of mammals and birds

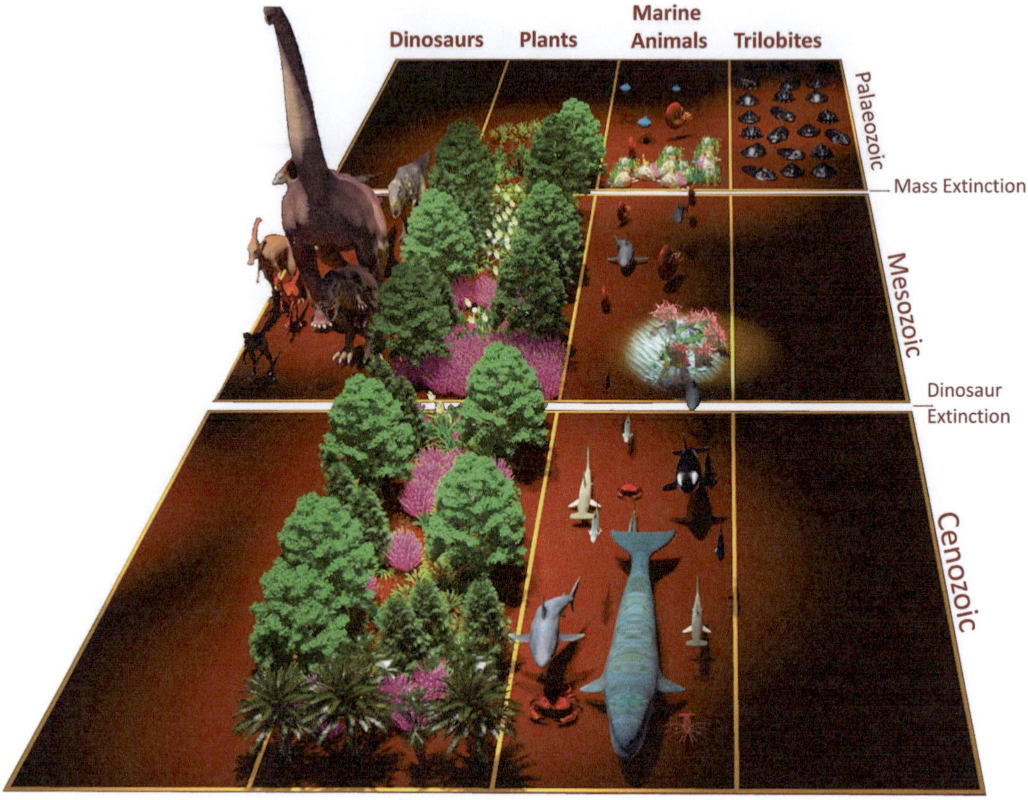

Fig. 4.3 A simplified overview of the main animal species in Oman during the eras of the Phanerozoic Eon (Palaeozoic, Mesozoic and Cenozoic). Plants evolved significantly during the Mesozoic Era, culminating in the appearance of flowering plants (angiosperms) during the Cretaceous which became predominant by the end of the period. The appearance of flowering plants was aided by the spread of insects, like bees, which are unfortunately often poorly fossilized. Marine life included ammonites, rudists, echinoderms, large fish and marine reptiles. The dinosaurs were the most prominent terrestrial fauna. Mammals also appeared in the Mesozoic, but no evidence for Mesozoic mammals has been found in Oman so far

appeared, and insects, like bees, spread out. This, in turn, supported the emergence and spread of flowering plants (angiosperms). Ferns and conifers were still the most prominent varieties of terrestrial plants, but new kinds appeared during the Cretaceous, along with whole new species like oaks and elms. Animals like crocodiles, turtles, frogs and big fish also spread around the globe.

After great prosperity and diversity, the number and type of organisms fell dramatically during the late part of the Cretaceous Period (Fig. 4.3). The boundary between the Mesozoic and Cenozoic, the Cretaceous–Palaeogene (K-Pg) boundary or Cretaceous–Tertiary (K-T) boundary saw another mass extinction, resulting in the complete disappearance of about three-quarters of animal and plant species. Dinosaurs, as well as many marine animals—particularly invertebrates—were lost forever. This extinction is often attributed to a great meteorite impact in the area of the Gulf of Mexico. It left behind a huge crater, 180 km in diameter, known as the Chicxulub Crater. The rocks of the impact were dated at 66 million years ago, coinciding with the end of Cretaceous. The rock and clay layers deposited during that period have a relatively high concentration of a rare element of the platinum family, known as iridium, which is present in higher concentration in meteorites compared to the earth's crust. It appears that the impact of the Chicxulub

Fig. 4.4 Mesozoic rocks form the great bulk of the mountains of Oman. This prominent mountain peak is known as Jabal Al Misht. It represents a spectacular oceanic seamount (mountain rising from the oceanic seafloor) that formed hundreds of kilometres offshore from the northern coast of Oman, in the Neo-tethys Ocean, at the end of the Permian and through the Mesozoic. The rocks exposed at the top are limestone from the Triassic Period, underlain by volcanic rocks from the same period. This oceanic mountain was emplaced on top of northern Oman at the end of the Cretaceous Period. During the process, the mountain was fissured by many faults and fractures. The cliff face of Jabal Al Misht is known to be the best rock climbing rock face in Arabia with a total height of more than 1000 m. This photograph was taken by Ken Glennie in the 1960s. Ken and his research group named these limestone mountains 'Oman Exotics'

meteorite was accompanied by hot dust, ash and rock fragments (including iridium). They covered the entire surface of the earth and blocked out the sunlight, plunging the world into semi-darkness. Temperatures fell, plants withered, and animals choked. The world became a harsh environment for animals—particularly dinosaurs—causing them to die of hunger, cold or asphyxiation.

4.2 Oman During the Mesozoic

During the Mesozoic Era, Oman had a unique location at the edge of the Tethys Ocean. This position enabled it to witness the many climatic changes and tectonic events of the period. A great proportion of the country was submerged under a shallow sea that covered many of the world's continents during the Mesozoic. Thick layers of carbonate rocks were deposited then. They could be seen in the mountain chains of Al Hajar Mountains, Al Huqf and the Dhofar Mountains (e.g. Fig. 4.4). Within these layers, there are a huge number of fossilized organisms that lived in the nutrient-rich waters of the shallow marine environments when Oman was positioned near the equator. These fossiliferous rocks form source (of about 15% of the oil) and reservoir rocks for hydrocarbons in Oman and the Arabian Gulf.

One of the most important Mesozoic geological events in Oman was the development of a mid-oceanic ridge parallel to the northern coast

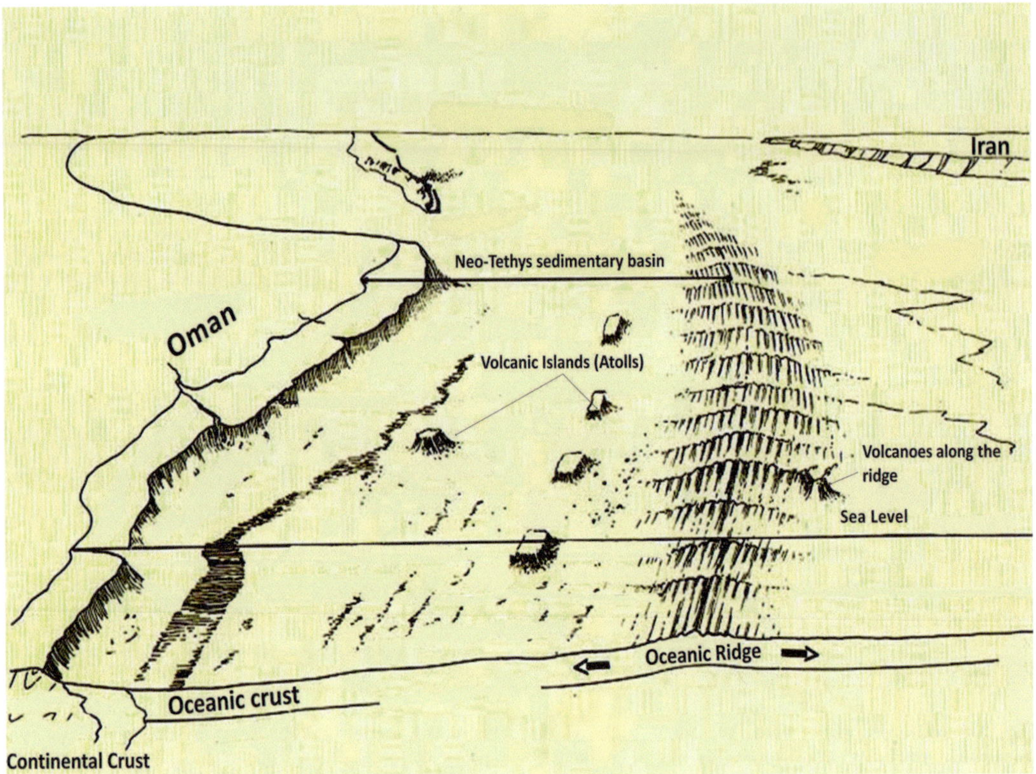

Fig. 4.5 Formation of a mid-oceanic ridge parallel to the northern coast of Oman to form a new ocean, known as the Neo-tethys Ocean as drawn by Samir Hanna in 1996. Magma rose along this ridge to form the Ophiolite rocks of Oman. This drawing shows the details of the volcanic islands and the oceanic deposits that formed within the abyssal plain (an underwater plain more than 3,000 m deep)

of Oman. Along this ridge, magma was rising, and Oman—with the rest of Arabia—was continuously separating from the northern micro-continents, such as Afghanistan and Iran, during the formation of a new oceanic crust (Fig. 4.5). On it, volcanic islands were formed, many covered by shallow seas which played host to marine organisms. In the deep oceans, chert—silicate sedimentary rocks—was laid down. During the Mesozoic, therefore, two different types of rocks were deposited in the ocean. The first were carbonate rocks above volcanic islands that formed parallel to the coast of Oman along the mid-oceanic ridge, and the second were silica-rich rocks at water depths of more than three kilometres. The deposition of these rocks continued for at least 150 million years from the beginning of the Mesozoic to about 90 million years ago.

At the end of the Cretaceous, South America separated from Africa, and the Atlantic Ocean started to expand along its mid-oceanic ridge, therefore pushing South America towards the southwest and Africa with Arabia towards the northeast. This process led to a gradual closure of the Neo-tethys Ocean on the northern side of Oman (Fig. 4.6). The oceanic crust of the Neo-tethys, including the deep oceanic sediments and the volcanic islands with their shallow carbonates, was obducted or pushed on top of the northern surface of Oman. The whole process took about 10 million years. During this period, the oceanic crust travelled hundreds of kilometres before it reached its final destination to settle on the surface of Oman,

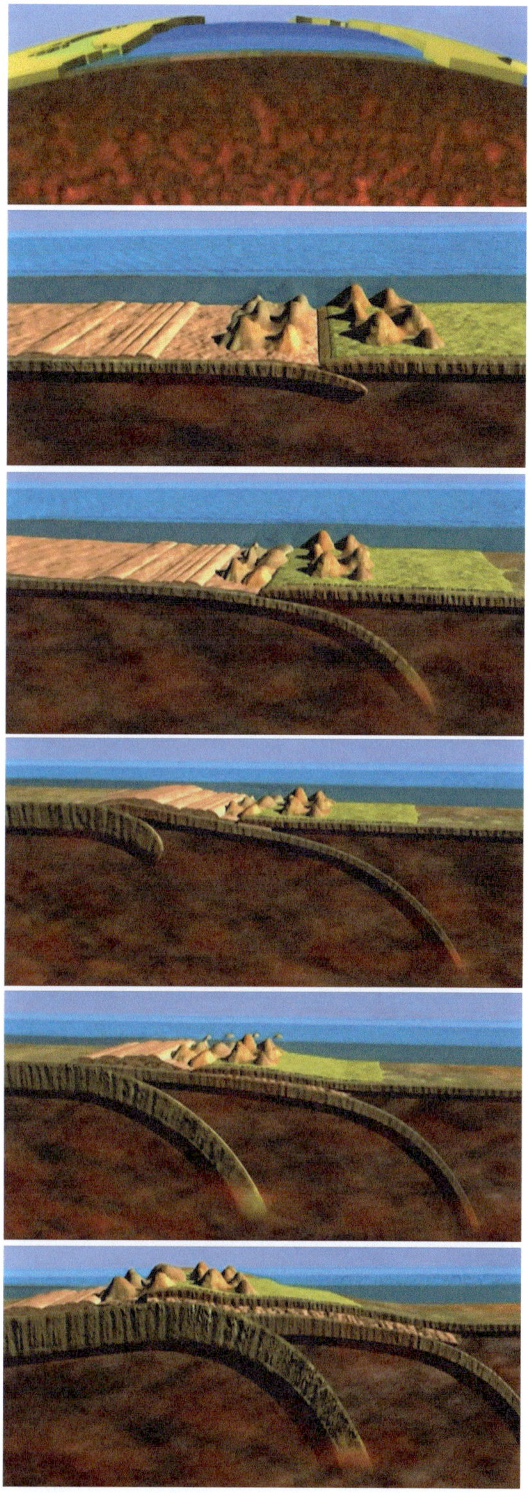

Fig. 4.6 Steps of the obduction process: **a** The first phase of obduction occurred in the middle part of a relatively vast ocean, known as the Neo-tethys, which is represented today by its small and closing remnant, the Sea of Oman. **b** The obduction process probably initiated along a pre-existing crustal crack or mid-oceanic ridge that existed between the northern side of Oman and the northern continents, along the Neo-tethys Ocean. **c** The third figure shows the movement of oceanic rocks towards the land of Oman and the transfer of deep oceanic sediments and volcanic islands with their overlying carbonate sediments (e.g. Jabal Al Kawr and Jabal Al Misht). **d** and **e**. Another subduction zone possibly developed between the continental plate of Oman (northern Arabia) and the attached oceanic crust, while the other oceanic crust continued to move towards Oman. The oceanic crusts with the Ophiolite and oceanic sediments were emplaced on top of Oman and the Arabian Plate subducted to great depths towards the mantle. **f** Rebound (isostatic uplift) of the Arabian Plate back upwards due to buoyancy is shown in the last figure (continental crusts have lower densities than the mantle and oceanic crusts.) Because of this rebound, the northern part of Oman emerged above sea level, therefore providing land for dinosaurs to live on

Fig. 4.7 As the oceanic sediments and seamounts moved from the ocean towards Oman during the Cretaceous Period, they suffered a lot of tectonics and deformation, causing them to fold and fracture. The oceanic sediments (pelagic sediments) are generally fine and form thin layers of beautiful colours, as shown in the top figure from Ras Al Hadd in the Wilayat of Sur. While being transported to land (the so-called obduction process), they formed tight folds that are often cut by compressional faults. The folds are often re-folded into complex structures. The lower photograph is from Tanuf in the Wilayat of Nizwa, and it shows limestone rocks, in the far and close distances, from the 'Oman Exotics' which were underlain and formed above volcanic islands

carrying with it many secrets about the life forms, geological processes and environmental changes in the deep oceans and along mid-oceanic ridges (Figs. 4.7, 4.8, 4.9 and 4.10).

During the latest stages of Ophiolite obduction and collision between the continental crust and oceanic crust (about 85 million years ago), the northern part of Oman was subducted or forced

Fig. 4.8 Various sequences from a mid-oceanic ridge of the Neo-tethys. On top are pillow lavas of Wadi Al Jizi in North Oman and in the picture below is sheeted-dyke complex. Sheeted dykes normally coalesce out from underlying gabbro layers that represent the lower part of the oceanic crust. The dykes open up and feed the overlying pillow lavas that extrude under oceanic water. Pillow lavas from the Neo-tethys Ocean are found in many places in Oman. The bottom photograph shows columns of basalt (mostly hexagonal) in the Village of Al Ghizayn in the Wilayat of Al Khaburah. These were formed during rapid cooling of thick lava flow that results in horizontal shrinking, producing contractional fractures that take this characteristic shape

Fig. 4.9 Top photograph is a view of Fanja, looking west. It shows the Al Khawd conglomerate on the right (black arrow) and the sharp and small dark peaks of Ophiolite. The yellow arrow indicates the line of Moho that separates the mantle rocks (dark colour) from the crustal rocks (pale colour). In the background is Al Jabal Al Akhdar. The bottom photograph is of the Ophiolite rocks in Ras Madrakah. This Ophiolite block is known as the Masirah Ophiolite. This very thin slab of oceanic crust from the Indian Ocean was emplaced on Oman during Late Cretaceous and Early Cenozoic

Fig. 4.10 On the right side a fossil black smoker in the Village of Al Ghizayn in the Wilayat of Al Khaburah and on the left side black smokers as they appear in the deep ocean (photograph by Chris German)

below the oceanic crust of the Neo-tethys and sunk deep into the mantle (refer to Fig. 4.6). While sinking, northern Oman reached depths in excess of 50 km, experiencing medium-to-high temperatures (around 540 °C) and high pressure (about 21 kbar). The limestone rocks that form the crust of northern Oman and were deposited in Late Palaeozoic and Mesozoic times underwent extreme metamorphism and significant alteration in these extreme conditions resulting in the change of limestone rocks to eclogite facies, which represent the highest grade metamorphism. These rocks are found along Al Sifah Beach in Muscat along a rocky shore exposure that extends for about 500 m (Fig. 4.11).

After the obduction process ceased, the forces that pushed Arabia beneath the oceanic crust of the Neo-tethys were relaxed, and the subducted continental crust of Arabia (in this case the northern part of Oman) bounced back to the surface in a process known as isostatic uplift of the lithosphere. This phenomenon is primarily caused by the light density of the continental crust compared to the mantle. As it rebounded, the rocks that were once forced down to depths of more than 50 km were raised, and the surface of the continental crust of northern Oman was exposed above sea level (see Fig. 4.6).

It is interesting to note here that the whole obduction process of Ophiolite probably occurred below sea level, and the northern part of Oman was only exposed above sea level during the isostatic rebound of the continent after obduction. While exposed above sea water around 75 million years ago and after, rivers flowed from mountains to sea, in a similar direction to the wadis in Al Khawd today (Fig. 4.12). As a result, the exposed peaks of Ophiolite and oceanic sediments were eroded and weathered, therefore forming thick successions of conglomerate gravels, known as the Al Khawd Conglomerate.

4.3 Organisms Living in Oman During the Mesozoic

4.3.1 Cartilage Fish

The shallow, marine limestone rocks of the Mesozoic are rich in fish fossils—especially those rocks at the top of Al Jabal Al Akhdar, at an elevation of about 2,000 metres. Far above sea level, is a rock bed consisting of tens of fossils of fish bones and teeth (Fig. 4.13). This rock bed was deposited at the boundary between the Jurassic and Cretaceous Periods. It marks the Jurassic uplift when the Arabian Plate was thrust upward and then subsided, drowning the land beneath and allowing the sea to creep in. The Cretaceous sea advancement or transgression is responsible for the deposition of the wealth of fossils which mark the Mesozoic layers. The fossils are mostly the teeth and bones of selachians such as sharks, rays and skates which

Fig. 4.11 Al Sifah Beach in the Governorate of Muscat hosts the eclogite rocks that represent the highest grade alteration or metamorphism to occur during the subduction of northern Oman below the oceanic crust of the Neo-tethys Ocean due to conditions of extreme pressure and high temperature. This location exhibits the deepest structural level or burial point of the continental crust of Arabia to the mantle during the last stages of Ophiolite obduction and collision with the oceanic crust of the Neo-tethys about 85 million years ago. As shown on the left, the eclogite mostly occurs as small boudins (stretched chunks of rocks) normally about 5–10 cm long

are all cartilaginous fish. Since their skeletons are mainly made of cartilage and not bone, they are usually not preserved in the fossil record. Their teeth, however, serve to identify them. Most are not then so different from today's species and have changed little in the last 150 million years.

4.3.2 Trace Fossils from the Deep Sea

Trace fossils are important geological indicators of the environment and depositional setting. They have often been used to indicate the depth of sea in which they were formed. This is

Fig. 4.12 Rock layers of the Al Khawd Conglomerate in the Wilayat of As Seeb, which were deposited by rivers, where the dinosaurs lived at the end of the Mesozoic. These layers are sandwiched between the Ophiolite rocks (the dark peaks in the lower part of the picture), which were then just obducted on top of Oman, and the white rock units of limestone that were deposited during the Early Cenozoic in shallow seas. Because the soil of Al Khawd Conglomerate is rich in minerals, it often gets covered by trees and wild herbs, as seen here. They can be well identified around Al Khawd and Fanja from their distinctive brown colour. The arrow represents the flow direction of the wadi from south to north, towards the sea, which is about 13 km north of this location, most likely in a similar direction to the rivers that existed here 70 million years ago; however, the shoreline was much closer then, as evidenced from rocks and fossils

because of their sensitivity to water depth, temperature, water energy, salinity and sedimentary setting. In general, vertical traces of organism form in shallow seas, and horizontal ones develop in deep seas. The study of trace fossils is known as ichnology, and a certain assemblage of trace fossils is known as ichnofacies or ichnofauna. For example, Skolithos ichnofacies, shown in Fig. 3.41, normally indicate shallow seas and tidal flats, whereas Cruziana traces are most likely produced in mid to distal sea shelves.

The Nereites ichnofacies represent a group of trace fossils that mostly form in deep-sea sedimentary fans along the base of slope of turbidity systems. They are thought to be a product of grazing trails of possible worms that live in deep seas and abyssal zones. Different forms of Nereites ichnofacies are identified, and many still need better classifications. One of the well-known components of the Nereites ichnofacies are the graphoglyptids, which form complex ornate forms of mostly horizontal meandering and net-shaped tunnels used for dwelling and to trap food. Their complexity and diversity apparently increased significantly from the Triassic to Cretaceous. During Late Cretaceous, graphoglyptids became suddenly widely diversified.

The deep-sea sediments of Oman contain exceptionally rich and highly diverse graphoglyptids trace fossils in the Late Triassic turbidite sediments. These sediments were emplaced on North Oman during the Late Cretaceous times, with the Ophiolite sequences. At least 15 types of graphoglyptids traces have been identified in a few locations on the southern and northern sides of the Al Hajar Mountains, in places like Wadi

Fig. 4.13 A layer filled with condensed bones and teeth from Al Jabal Al Akhdar in the Wilayat of Nizwa at an elevation of about 2,000 m above sea level. This layer marks the boundary between the Jurassic and Cretaceous Periods and is, therefore, the lowermost bed of the Cretaceous strata. It was deposited following a significant subsidence and drowning of the platform at the beginning of Cretaceous when this area of Al Jabal Al Akhdar was overlain with thick column of sea water. This was responsible for the deposition of very fine sediments, cherts and abundant sea fossils. The hollow in the central bone (where the arrow is) could be for the spinal cord. The inset picture in the bottom left corner shows fossil bones with visible bone tissues, each fossil is about 2–4 cm long

Hibi, Wadi Saal and Al Jil. Two types of these exhibit new features, different from previously described graphoglyptids, and hence form two new ichnogenera of graphoglyptids. Together, these fossils form important collections from the Triassic and have significantly helped to understand the diversity and climate conditions during this time period (Fig. 4.14).

4.3.3 Belemnites

The genus Belemnitida or Belemnites are part of the Mollusca phylum and Cephalopoda class, and their fossils are common in the Mesozoic carbonate rocks of Oman. Belemnites are known for their bullet-shaped shells, pointed from one end and cone-shaped from the other (Fig. 4.15). In fact, their name is derived from the Greek belemnon meaning 'dart'. Generally, they are similar to squid, having ten arms, but squid lack an internal skeleton and have two specialized arms. The arms and fins of belemnites are rarely preserved in rocks, whereas their calcite or aragonite shells are often well preserved. They first appeared during the Jurassic, became common in the Early and mid-Cretaceous and died out at the end of the Cretaceous, making it an entirely Mesozoic creature. However, belemnite fossils have been found in some Cenozoic rocks around the world, including the rocks of Muscat in Oman. It is unclear whether these were belemnites that actually lived in the Cenozoic (a straggler that survived the extinction) or they were re-deposited from the weathered and eroded debris of Mesozoic rocks.

4.3.4 Ammonites

Ammonites are one of the most well-known marine fossils from the Palaeozoic and Mesozoic Eras. Their fossils are probably among the first ones recognized and utilized by scientists since they are excellent index fossils and may be used to determine the age of rocks. They also provide a lot of information about palaeoclimates.

Many myths and legends were introduced by ancient civilizations and cultures around the world like the Greeks and Romans to explain their presence in rocks. For instance, in England, ammonites were called 'snake stones', because people believed they were snakes, which curled up and were turned to stone by the devout prayers of Saint Hilda, around 650 A.D. In fact, there is a tradition of carving snake heads on ammonite shells, and there is actually a beautiful ammonite genus named Hildoceras, after St. Hilda.

The name is derived from Ammon, the ancient Egyptian god, who was always represented wearing ram's horns which, ribbed and coiled, mirror the shells of ammonites. Ammonoidea belong to the Cephalopoda class of the Mollusca phylum. The septal surfaces of ammonite shells with their bulges and indentations often represent different internal chambers (Fig. 4.16). They had a strand of tissue, known as the siphuncle, passing along the shell and connecting all the chambers. The siphuncle often distinguishes them from living nautiloids. It is possible that they originated from orthoconic cephalopods—which also had siphuncles—during the Devonian Period in the middle part of the Palaeozoic.

Some ammonite species died out during the mass extinction at the end of the Palaeozoic, but many survived to the Mesozoic. They became both diverse and widespread during the Jurassic and Cretaceous Periods, but were entirely wiped out during the mass extinction at the end of the Mesozoic. Their fossils are present in the Palaeozoic and Mesozoic rocks of Oman, but they are particularly abundant in Cretaceous rocks (Figs. 4.17 and 4.18). They are found on Al Jabal Al Akhdar in Nizwa, the Salakh and Madmar mountains in Adam, Jabal Qusaibah in Bahla and in the wilayats of Ibra and Ad Duqm.

4.3.5 Rudists

A site in the Ramlet Ghunaim in the Wilayat of Muhut is considered one of the best for rudist fossils among the marine fossil collections of the Mesozoic. They are more than 70 million years old and provide a record of the shallow, open seas at the end of Cretaceous when most manner of sea creatures lived in massive reefs where dissolved oxygen and nutrition were abundant. These reefs had large quantities of calcium carbonate required by some animals—including rudists and corals—to build shells and skeletons. Their name comes from the Latin rudis meaning 'primitive'. Rudists are molluscs that belong to the class Bivalvia.

Rudists first appeared during the Jurassic Period and became widespread during the Cretaceous. As with many other molluscs, they were completely wiped out during the mass extinction of the Late Cretaceous. Rudists filled large reefs in the shallow seas of the Tethys Ocean. Some upright types probably had tentacles that were used to filter food from the water (Fig. 4.19). They often had a conical shape, where the pointed end was attached to the sea floor or to other neighbouring rudists.

At least four different types of rudist shells can be identified in the Ramlet Ghunaim. Many fossil corals, other types of bivalves and gastropods are scattered among them (Fig. 4.20). However, more than 90% of the fossils in this area belong to a rudist species known as Vaccinites. It seems that these were more successful on the Cretaceous reefs of Muhut. These reefs were apparently very rich in terrigenous sediments (those derived from erosion of rocks and transported to sea) that supported the thriving of rudist reefs and controlled their distribution.

Fig. 4.14 **a** Omanichnus virgatus (right) and Vitichnus nizwaensis (left) are two new graphoglyptid ichnogenera found in the Upper Triassic turbiditic deep-sea fan deposits in Oman and named after Oman and Nizwa (Wetzel et al. 2009). **b** Various types of deep-sea burrows, including what could possibly be the following types: Paleodictyon (top left), Desmograpton (top right and central left (?)), Saerichnites (central right), Gordia (bottom left) and Imponoglyphus. The diversity of deep-sea ichnofacies from Oman during the Early Mesozoic (Late Triassic) was probably supported by the palaeoposition of Oman during that time near the equator. The length of each burrow trace is from 5 to 15 cm

Fig. 4.15 Left: a reconstruction of the possible shape of the extinct belemnites, showing a squid-like animal. Below: a cluster of many belemnite fossils with their bullet shape, from the base Cretaceous beds in Al Jabal Al Akhdar. Fish bones are present between the belemnite fossils. The fossils range in size from 3 cm to about 15 cm. Belemnites appeared in the Lower Jurassic and became extinct at the Upper Cretaceous. The fossils of belemnites are widespread in the Cretaceous limestone beds of north and South Oman

Fig. 4.16 Reconstruction of an ammonite. Ammonites are thought to be close relatives to animals like squid and cuttlefish and very similar to nautilus. They first appeared in the Devonian Period (about 400 million years ago) in the Palaeozoic Era and became extinct along with dinosaurs at the end of the Mesozoic, around 66 million years ago. Ammonites evolved into many different forms and species and so are considered to be excellent index fossils to correlate and identify the age of rocks

Rudist fossils are found extensively in the wadis of North Oman, such as Wadi Al Sahtan, Wadi Bani Kharus, Wadi Bani Awf and Wadi Al Muaydin (Fig. 4.21). They can also be found in many places in Dhofar. Many genera and species are endemic to Oman. The Late Cretaceous rocks of Oman are exceptionally rich and diverse in terms of rudist fossils. They provide a great deal of information about the evolutionary trends of rudists and the palaeoclimate around the Neo-tethys Ocean. They also form a large part of the hydrocarbon reservoir rocks of the subsurface of Oman. These have high porosity and permeability, enabling the storage and production of large quantities of oil and gas. Most belong to the Cretaceous Period and make up many of the large oil and gas fields like Qarn Alam, Jibal, Lekhwair and Fahud in the northern part of Oman. However, rudist reefs are susceptible to alteration, dissolution and recrystallization, causing variable degrees of deterioration in the quality of their petroleum reservoirs.

4.3.6 Megalodon Bivalves

The carbonate beds of the atolls that developed above volcanic islands of the Neo-tethys Ocean are characterized by the presence of a type of bivalve from the genus Megalodon (Fig. 4.22). These bivalves appeared in the Devonian and became extinct in the Jurassic. The Jurassic relatives of Megalodons are thought to be related to rudists, but most of the fossils in Oman belong to the Triassic Period. They are widespread in many Wilayats of northern Oman, such Bahla, Ibri and Sohar. Some can also be seen in manufactured slabs of limestone, which are often called Oman Marble. Along with bivalves, these slabs frequently include several types of coral, foraminifera and gastropods from the Mesozoic Era. Generally, the 'Exotics' are among the most fossiliferous rocks of the Mesozoic and Late Palaeozoic.

4.3.7 Gastropods

Gastropods or snails first emerged in the seas of the Late Cambrian and continued to diversify throughout all the periods of the Palaeozoic. By the Carboniferous Period, they had migrated from the seas and begun to live on land. Compared to other molluscs, like bivalves, the Gastropoda class was less diverse during the Palaeozoic, and their species were more primitive. However, with the beginning of the Mesozoic, they increased and spread out significantly in both fresh and sea water. Their shells became more coiled with time, and by the Cenozoic, they had become broadly similar. Fossil shells are widespread in the marine and terrestrial sedimentary rocks of the Cenozoic. Overall, an estimated 600 families of gastropods have appeared since the Palaeozoic, 200 of which disappeared during the previous extinction events. The families divide to about 80,000 species that live in all kinds of environments across the globe. The fossils of gastropods are usually well-preserved throughout Oman in the rocks of the Palaeozoic, Mesozoic and Cenozoic (Figs. 4.23 and 4.24).

Fig. 4.17 Fossils of ammonites are common in the Late Palaeozoic and Mesozoic limestone rocks of Oman. Because they appear in many different forms, they are often used as index fossils to date rocks. Ammonites found in Oman have different sizes and types of septa (the walls that divide the chambers), and they exhibit various degrees of coiling of the shell. The top photograph is of Early Triassic ammonites in a red limestone unit from the 'Oman Exotics' of the northern Oman mountains. These ammonites lived in isolated seamounts in the Neo-tethys Ocean. They are highly diversified and well preserved. The size of each ammonite is about 5 cm across. Photograph is provided by Dr. Aymon Baud. The bottom photograph shows larger fossils from North Oman. They are about 20 cm across and have beautiful spirals

Fig. 4.18 Top: Fossil of a new species of ammonite, called Baidites hermanni. The name is after Al Baid Village in the Wilayat of Dama Wa At Taiyyin, where the fossil is found in the 'Oman Exotics' of Early Triassic rocks. Bruehwiler et al. (2012) also reported the presence of one new ammonite family Galfettitidae, five new genera Baidites, Goundemandites, Lucasites, Omanites and Safraites and 12 new species among which are Baidites hermanni, Omanites musjahensis, Safraites simplex and Rohillites omanensis. These names are mostly derived from nearby villages or wadis. This highly diversified, and abundant fauna of ammonites indicates that they recovered quickly after the end of the Permian mass extinction, when more than 90% of all marine species became extinct. They provide valuable data about the distribution of ammonites in the Tethys Ocean. Nearly, all known Early Triassic ammonoid faunal assemblages are present in the exotic blocks from Oman. The bottom fossil is for another type of ammonite known as Galfettites simplicitatis from Jabal Safra, east of the Wilayat of Adam

4.3.8 Dinoflagellates

Dinoflagellata represent a phylum of the Protista kingdom that live as plankton in marine environments or, less prominently, in freshwater and ice. Their distribution is directly related to the depth, temperature and salinity of the water in which they live. Their name is derived from the Greek dinos meaning whirling and Latin flagellum meaning whip. Dinoflagellates are quite similar to plants and are divided into many different species that lived and continue to live in vast numbers since their fist appearance in the Silurian Period of the Palaeozoic Era. While their numbers were limited during the Palaeozoic, they became widespread at the end of the Mesozoic Era. Today, there are about 1,700 marine species and about 200 freshwater species, although these

Fig. 4.19 Above: a possible reconstruction of rudist and coral reefs in shallow seas and below is the source bed of the rudists in Ramlat Ghinayyim. The bed is about three metres thick, and it is nearly completely filled with rudists

numbers are rapidly increasing as scientists continue to make new discoveries. Dinoflagellates are mixotroph organisms which combine different sources of energy and food. They produce energy through photosynthesis or by preying on other marine organisms including fish, taking full advantage of their environmental circumstances. Concentrations may reach a million cells in one millimetre of water, resulting in large quantities of toxic excretions which cause harmful algal blooms, such as the 'red tide'. This affects other marine animals and the people who consume seafood (Fig. 4.25). Such blooms are directly related to concentrations of phosphates and nitrates in water as a result of human activity.

Some types of dinoflagellates produce calcium carbonate cysts as part of their life cycle. The fossils of these cysts are common in rocks of all time periods of the Mesozoic and Cenozoic in Oman (Fig. 4.26). They are used as index fossils to date rocks as well as being used to understand climatic and biological changes during these periods.

Fig. 4.20 Different rudist fossils from Ramlet Ghunaim in the Wilayat of Muhut. The top picture is a biostrome, where fossils of a certain type of rudist known as Durania have been exceptionally well-preserved in an almost life-like position. The fossils of Durania are generally larger than other types of rudists and have large coiled valves. The middle picture shows a Vaccinite rudist with its top structure nicely preserved. Vaccinite rudists are by far the most common in Ramlet Ghunaim. The bottom picture shows another type of rudist known as Torreites. It looks like a human hand, but with only four fingers. Some types of Torreites are known as Praetorreites omanesis. Many taxa of rudists found in Ramlet Ghunaim are endemic to Oman. These rudists thrived in the warm, shallow waters of the Tethys ocean. Overall, some 70 different species and 40 genera of rudists have been identified in Oman

Fig. 4.21 Rock units filled with different types of rudists in Al Jabal Al Akhdar (top) and Jabal Al Sumayni in the Wilayat of Mahadah (bottom). The fossils belong to the Cretaceous Period and are estimated to be 100 million years old in Al Jabal Al Akhdar and about 70 million years old in Jabal Al Sumayni, similar to the fossils in Ramlet Ghunaim. Jabal Al Sumayni is considered the second best site in Oman to see rich, diverse and well-preserved rudist bivalves. An individual bivalve is normally around 10 cm long

Fig. 4.22 Fossils of Megalodon bivalves from the Mesozoic seamounts of the Neo-tethys ocean. The top picture is a limestone bed full of Megalodon bivalves, and the bottom picture is a complete Megalodon fossil

Fig. 4.23 Different types of gastropod or snail fossils from the Mesozoic carbonate rocks of Oman. Top: well-preserved gastropods from Late Cretaceous in the Wilayat of Mahadah. Bottom: gastropod fossils from the Jurassic rocks of Al Jabal Al Akhdar. Many gastropod species, including various species of the Turritella genus, appeared in the Cretaceous Period

Fig. 4.24 Mesozoic rocks in Jabal Harim, the highest peak of Musandam (about 1800 m above sea level) with large gastropod fossils and possible fish bones. The inset picture shows a possible fish fossil and a gastropod, of about 15 cm long

4.3.9 Dinosaurs

Dinosaurs are probably the most prominent life form of the Mesozoic. The largest animals that ever lived on land, and they are one of the most diverse groups of vertebrates, in terms of shape, size and behaviour. However, the name 'dinosaurs' is misleading. It comes from the Greek dienos meaning 'terrible' or 'wondrous' and saurus meaning 'lizard'. In fact, dinosaurs are not related to lizards, and many were quite small. They first appeared at the end of Triassic and were completely wiped out during the mass extinction at the end of the Mesozoic—a period of some 135 million years. During that time, at least 1,000 species of non-avian dinosaurs appeared. They are often classified into two main groups based on the structure of their hip bones—reptile-hipped or bird-hipped. Reptile-hipped dinosaurs had a more primitive hip structure.

Dinosaur bones have been found in various locations in the vicinity Al Khawd in the Wilayat of As Seeb (Fig. 4.27). This is one of the best places in the Arabian Peninsula to find them. During the Cretaceous Period, the area was criss-crossed by rivers and streams, the climate was much wetter than today and there was far more frequent rainfall. These Cretaceous rivers flowed from the mountain heights down towards the sea—in a similar direction of flow to existing wadis (see Fig. 4.12)—carrying with them animal bones and wood fragments which were re-deposited in estuaries, where the rivers met the sea. This probably explains why no complete skeletons of dinosaurs have yet been found here.

At least four types of dinosaur have been identified in Oman. The first species belong to the small herbivorous Ornithopoda—running grazers that dominated the land during the Cretaceous after developing a sophisticated chewing

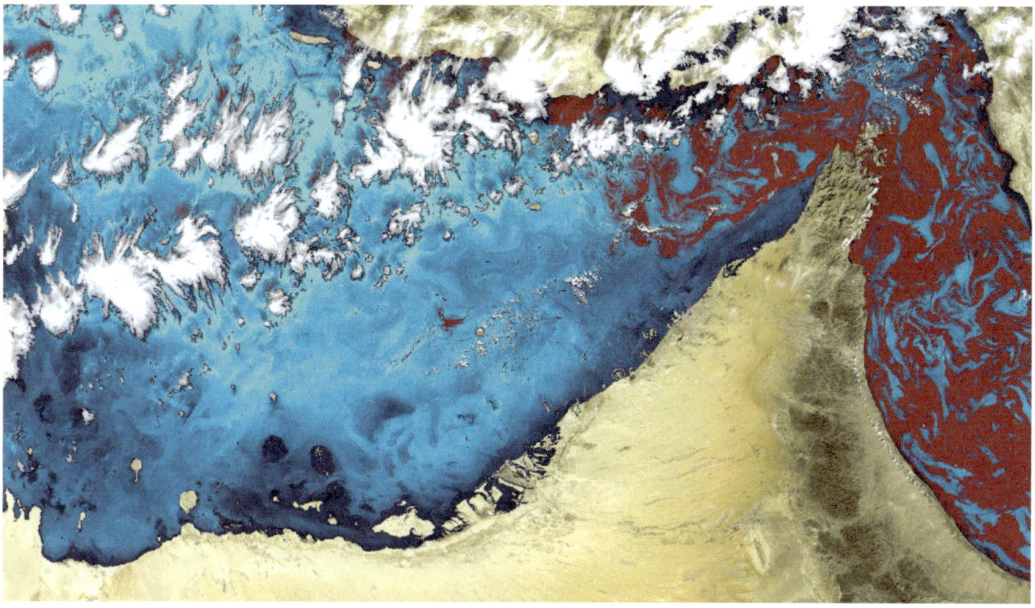

Fig. 4.25 A satellite image by ESA (European Space Agency) acquired by Envisat's MERIS instrument on 22 November 2008 showing the red tides off the coast of Oman and UAE. This harmful algal bloom is caused by a huge concentration of microscopic algae, mainly dinoflagellates, resulting in the red or pink coloration of the surface water. Such blooms pose a threat to marine life and ecosystems and can cause serious harm to human health

Fig. 4.26 Fossils of dinoflagellates from the Mesozoic Era from Petroleum Development Oman. Fossils are widespread and common in Omani waters. Living dinoflagellates are also diverse and prevalent in the seas of Oman, but large numbers cause disease among marine organisms and humans

Fig. 4.27 Representation of the four types of dinosaur found in Oman. From left to right, these are the clades Ornithopoda, Theropoda, Sauropoda and Hadrosauridae showing approximate relative sizes. The largest dinosaur, sauropod, might have been more than 30 m long

apparatus, which gave them an evolutionary advantage compared to other dinosaurs (Fig. 4.28). Their name comes from the Greek ornithos meaning 'bird' and poda meaning 'feet' since they had three-toed feet similar to birds' feet. Ornithopods were up to fifteen metres long and weighed about twenty tons—bigger than a bus. The second type belonged to the carnivorous Theropoda whose name in Greek comes from ther meaning 'beast' and poda meaning 'feet' (Fig. 4.29). They probably appeared during the Triassic and became the sole carnivores during the Jurassic and Cretaceous. Therapods ranged widely in size from the size of a chicken to the size of a house. The third species found in Oman comes from the giant Sauropoda—the largest ever animals to live on earth. Their name means 'lizard-footed' from the Greek, sauros for 'lizard', and they were probably the most widespread dinosaurs during the Jurassic. They were certainly the largest at more than thirty metres long and eighteen metres tall—the size of a five-storey apartment block (Fig. 4.30). The final and most recently discovered dinosaurs in Oman are the herbivorous duck-billed dinosaurs of the genus Hadrosauridae (hadros means 'sturdy' in Greek) (Fig. 4.31). It was previously believed that this dinosaur only existed in Europe, East Asia and North America. Various fragments of dinosaur bones are still not well identified from Oman (Fig. 4.32).

During the geological survey of the Al Khawd Area in 2014, the Geological Society of Oman's team found a number of possible gizzard stones from the intestinal tracts of animals like dinosaurs. They were used to grind food particularly by animals that lacked grinding teeth (e.g. some types of herbivorous dinosaurs, like sauropods). These rocks are often known as gastroliths (gastro meaning stomach in Greek) and are often polished and well rounded (Fig. 4.33). They sometimes include long, hairline scratches formed by contact with the sharp edges of freshly swallowed stones. They are extremely hard to identify as it is difficult to know whether a particular polished clast is a gastrolith or not. Their

Fig. 4.28 Rhabdodon Ornithopod like the one found within the Cretaceous conglomerate of Al Khawd deposited in a brackish-freshwater aquatic environment. Rhabdodons lived during Late Cretaceous, approximately 66–70 million years ago. They were important herbivores during that period. The fossils of Rhabdodon found in Oman include different vertebrates. These dinosaurs were only about 3 m long—small dinosaurs that might have been prey for other carnivorous dinosaurs like theropod

Fig. 4.29 Carcharodontosaurus theropod, like the one found in Oman, represent large and heavy carnivorous dinosaurs that might have reached up to 10 m in length and had long teeth of around 20 cm in length. Its name in Greek means 'sharp teeth'. It probably lived in the vicinity of Al Khawd Village about 70 million years ago

interpretation was only confirmed worldwide about a century ago when they were found inside sauropod remains. They vary in size from sand to cobbles, depending on the size of the dinosaur and its diet. Some gastroliths may have been used by aquatic animals, such as plesiosaurs, to help them dive inside water and balance their buoyancy.

Among the fossil traces of dinosaurs found around Al Khawd Village are fossilized faeces, known as coprolite (Fig. 4.34). The name is derived from the Greek kopros meaning 'dung'.

Fig. 4.30 Rapetosaurus possibly lived in Al Khawd about 70 million years ago. These herbivorous dinosaurs might have reached a length of 15 m and had long neck and slender tail. Compared to other sauropods, including titanosaurs, Rapetosaurus were modest in size, but would have weighed as much as an elephant. They lived during the Late Cretaceous and were among the last dinosaurs roaming the earth before the mass extinction. They were first found in Madagascar, hence the name 'Rapeto' which is derived from a giant in Malagasy folklore

Fig. 4.31 Fossils of Hadrosaurus ornithopods were recently found near Al Khawd Village. These dinosaurs were common during Late Cretaceous, particularly in Asia, Europe and North America. It was previously thought that they were limited to one area of Gondwana; however, the recent discovery might reveal that Hadrosaurus was able to cross the Tethys Ocean either via small lined up islands or through a previously unknown link between Gondwana and Laurasia

Coprolites come in various shapes depending on the size and diet of the dinosaur. They can be identified from their external morphology and the remains of plants or bones inside. They provide a lot of information about the types of dinosaurs (herbivorous or carnivorous) that lived in a certain area and the food available to them. Coprolites have also been used as fertilizer because of their high phosphate content. However, much of the original composition of coprolites is often replaced by other minerals, so they lose their primary organic content. Al Khawd possibly also has crocodile coprolites. These are often smaller than dinosaur ones and have finer, chalkier matrix.

Dinosaur bones and teeth have recently been found in the Qahlah Formation in the Governorate of Al Buraimi, in northwest Oman. Just as the Al Khawd Conglomerate, the Qahlah was deposited following the emplacement and subsequent weathering of the Ophiolite and associated oceanic sediments on North Oman during

Fig. 4.32 A collection of dinosaur fossils found recently in Oman mostly by Axel Hartman and the author. They range in size from about 55 cm long to about 10 cm long

Fig. 4.32 (continued)

Fig. 4.33 Sauropod eating habits. On the left, an artistic reconstruction of a sauropod eating fronds of cycads and plant roots along with gastrolith pebbles which help it to grind the food (original drawing by Gordon Malon). On the top right, two samples—a possible gastrolith on the left and a normal rock on the right. Gastroliths are distinguished by their highly polished surfaces and tiny straight striations, usually along the long axis of the rock, formed as they scratched against each other inside the dinosaur. On the bottom right, a gastrolith with a polished surface and small scratches from top to base. Both samples were found by Dr. Martin Pickford during the survey work of the Geological Society of Oman in the vicinity of the old village of Al Khawd in 2014

the Late Cretaceous. This sequence, 100 s of metres thick, was the first deposited above the obducted plates in mostly alluvial and fluvial environments.

The new finding in the Qahlah Formation in Al Buraimi included a tooth of what could well be a mosasaurus tooth (Fig. 4.35). Mosasaurs were carnivores, famous to be the T-rex of the sea. They lived in about the last 25 million years of the Cretaceous, from about 92 million years ago to 66 million years before disappearing at the Late Cretaceous mass extinction. The stabbing rounded teeth of mosasaurs allowed them to hunt and swallow big animals, like sharks, and crush almost all animals with hard shells, like ammonites and other molluscs; therefore, no other rivals probably competed with them in the seas during their existence. Mosasaurs have most likely breathed air and had flexible, powerful and long bodies that allowed them to become excellent giant swimmers. The evolution history of mosasaurs remains uncertain. They have probably evolved from semi-aquatic lizards during the Early Cretaceous. They flourished and diversified because of the high nutrient delivery during the Cretaceous. Mosasaurs varied quite significantly in size from less than a metre to probably more than 15 m length.

Fossils of mosasaurs have been found in many places around the globe, with many different genera and species; however, none have been found in Oman until recently. The recent find of a mosasaurus tooth is the first identification of a swimming dinosaur in Oman. It is also the first find of a dinosaur tooth. The tooth is about 4 cm long and 2 cm wide. It was found in the weathered debris of the Qahlah Formation. The tooth was probably weathered from fan-delta rock facies deposited in shallow marine conditions. The exposures of the Qahlah Formation in the

(a)

(b)

Fig. 4.34 A possible dinosaur coprolite from Al Khawd. Remains of plants can be seen within this trace fossil, indicating that it was most likely produced by a plant-eating dinosaur. The fossil was found by Dr. Martin Pickford. Below, the brown outcrops of the Al Khawd Conglomerate are mainly found along small gullies, where the overlying scree of the white Cenozoic limestone is weathered out. These outcrops most likely still contain a lot of information about the behaviour of dinosaurs during the Late Cretaceous in Oman

area also includes fossil wood and leaves within what looks like pebbly sandstone facies deposited probably in braided streams.

4.3.10 Cretaceous Trees

Along with dinosaur bones, many petrified tree fossils have been found in the Cretaceous conglomerates of Al Khawd and Fanja in the wilayats of As Seeb and Bidbid. These trees were a food source for herbivorous dinosaurs, which were themselves prey to the carnivorous dinosaurs. They were probably gymnosperms, such as conifers, ferns and cycads. Some of the plant fossils in the area have been classified as palm trees. If confirmed, at more than 70 million years old, they will be the oldest known palm tree fossils in Arabia (Figs. 4.35, 4.36 and 4.37).

Fig. 4.35 **a** A possible reconstruction of the giant marine mosasaur (Kansas Geological Survey, Educational Series 6, Evans,1988). Our understanding of the shape of these dinosaurs have changed recently. **b** Two views of the mosasaur tooth from the Governorate of Al Buraimi, in Northwest Oman. The tooth is about 4 cm long and 2 cm wide. **c** The area also contains fragments of petrified wood from Late Cretaceous. **d** The The bottom photograph shows the red-coloured outcrops of Qahlah in Al Buraimi

Fig. 4.36 Fossil tree trunks in Al Khawd (on top) and the Wilayat of Dama Wa At Taiyyin (below) from the Cretaceous Period. The lower trunk is likely to be for a palm tree. It is about 70 cm long

4.3.11 Lung Fish

A number of marine fossils are found in Al Khawd area, including bivalves, gastropods and foraminifera. One of the most interesting of these is the fossil of a lung fish known as Ceratodus (Fig. 4.38).

4.3.12 Crocodiles

A number of fossils of the bones and teeth of crocodiles that lived during the Mesozoic have been found in the Al Khawd area (Fig. 4.39). There are also samples of turtle shells which look very similar to the ones found today along the

Fig. 4.37 Fossils of possible roots, from the Late Cretaceous. They look similar to the roots of palm trees today. They may be among the oldest palm trees and flowering plants that appeared in the world during the Cretaceous Period. There are more than 200 genera and 2,600 species of palm tree known worldwide, most of which grow in tropical and warm climates. They may inhabit rainforests and deserts. Recent palm tree roots from Oman are shown in Fig. 4.37

Fig. 4.38 On the top left are the roots and leaves of a living palm tree in Oman. On the top right is a fossil leaf from the Late Cretaceous in the Wilayat of Dama Wa At Taiyyin. It looks similar to modern leaves. Below are different fossil leaves from the same area and the same age. Most of these are angiosperms (flowering plants). The shape of the leaves indicates a wet climate

Fig. 4.39 A lung fish, most likely of the extinct genus Ceratodus (meaning 'horned tooth' in Greek). The illustration on the left is by Heinrich Harder, and the fossil samples on the right are an internal view of the tooth which is about 4 cm long, found by Martin Pickford

Fig. 4.40 Top figures are fossils of crocodiles from Al Khawd dated to around 70 million years ago. Top left: a crocodile tooth (hammer for scale) found by Axel Hartman. Top right: a bone found by Martin Pickford, about 10 cm long. Below are crocodile fossils from the Cenozoic rocks of central and southern Oman. On the bottom left, a crocodile jaw about 15 cm long and on the right a crocodile vertebra about 10 cm long

coast of Oman. A large group of bivalve and snail fossils, as well as fish teeth, lie within rock layers. All these animals lived alongside the dinosaurs, above and inside a Cretaceous estuary environment in Al Khawd, at a time when shallow seas covered large parts of land, and many animals, particularly dinosaurs, were competing to find suitable land to live on.

The Late Cretaceous fossil remains of crocodiles in Al Khawd represent the oldest evidence for these animals in Oman. They most likely appeared during the Early Mesozoic. They were small, terrestrial, and some were plausibly vegetarian. By the Cretaceous, they had abandoned their terrestrial lifestyle and developed in size—some to 10 m in length, like the famous Sarcosuchus crocodile. However, these enormous creatures, like their dinosaur cousins, did not survive the mass extinction at the end of Cretaceous. Modern crocodiles have evolved very little since the Late Cretaceous.

4.3.13 Turtles

Like crocodiles, turtles most likely evolved in the Early Mesozoic (Triassic Period). The oldest known fossil evidence in Oman belongs to the Cretaceous Period (Fig. 4.40). The conglomerate of Al Khawd is full of fossil remains of Cretaceous turtles. Most of these fossils look very similar to bone remains of turtles found along beaches today. Some of the Cretaceous turtles may have been up to 4 m long, but similar to giant crocodiles, they probably did not survive the mass extinction of the end of the Cretaceous. Overall, however, the aquatic lifestyle and slow metabolism of most turtle species helped them not only survive, but be barely affected by the extinction event. Many only died out during dry events in the Early Cenozoic.

Fossil turtles are very common in the Cenozoic rocks of central and southern Oman, particularly in Ghabah and Aidum. Further studies

Fig. 4.41 Turtle fossils. Remarkably, similar turtle remains. Left—Cretaceous, found by Axel Hartman, right—a recently dead specimen from As Sifah beach. The samples are from 15 to 20 cm long. Below: a bed of Al Khawd conglomerate with a bone that possibly belongs to a Cretaceous turtle

of remains in Oman may reveal some of the evolutionary processes of sea turtles.

4.3.14 Summary of Life in Oman During the Mesozoic

Life in the Mesozoic and the environment in which it flourished are represented in the image on the following page, summed up by the end of the Cretaceous in the Al Khawd area (Fig. 4.41), as drawn by Oscar Sanisidro for this book. Following the Ophiolite obduction of the Late Cretaceous, the area was uplifted above sea level. Fossil remains of dinosaurs in Al Khawd are represented by a large-sized theropod (Carcharodontidae), a small ornithopod (Rhabdodon-like) and a medium-sized sauropod (Paralititan-like). In the background on the left are a number of Hadrosaurus ornithopods. These dinosaurs represent the main types discovered so far in Al Khawd. Trace fossils of dinosaurs in the area included coprolites and gastrolith pebbles.

At the end of the Cretaceous, Al Khawd was then most likely an estuary, with one or many rivers flowing from south to north, in a similar direction to the flow today, and a body of brackish water connecting them to the open sea. The presence of terrestrial, brackish water and marine fossils, including dinosaurs, crocodiles,

Fig. 4.42 Summary of life in the Mesozoic, represented by the Late Cretaceous fauna and flora in Al Khawd, showing four different types of dinosaurs, crocodiles, turtles and various trees, including flowering ones. The drawing is made by Oscar Sanisidro for this book. The background mountains are for the Ophiolite mountains of Al Khawd, which had already been obducted in the area at that time (about 70 million years ago)

turtles, bivalves, gastropods and foraminifera, indicates that the area formed a transition zone between river and maritime environments.

Fossils of crocodiles and turtles are quite abundant in Al Khawd, although they are not well identified for their specific taxon. Crocodilians are represented by an undetermined Trematochampsidae species (similar to Trematochampsa) and a longirostrine crocodile. There are also some turtles.

Unfortunately, the details of floral remains are also not well identified yet. Some Late Cretaceous floras from Gerofit and Qetura, Negev Desert (Krassilov 2005), have been used to reconstruct the floral environment. It includes the genera Platydebeya, Menispermites (small trees/bushes), Dewalquea, Gerofitia and Gerocladus.

Beyond the details described in Fig. 4.42, Oman had a unique location at the edge of the Tethys Ocean during the Triassic, Jurassic and most of the Cretaceous. The land of Oman was covered by shallow, tropical sea water that hosted different types of fish, molluscs and corals. Shallow reefs had large quantities of calcium carbonate required by some animals—including molluscs and corals—to build shells and skeletons. Among the Mesozoic marine fossils found in Oman are selachian fish that comprise sharks, skates and rays. Oman also includes a unique collection of Early Triassic ammonites. These provide valuable data about the distribution of ammonites in the Tethys Ocean. Nearly, all known Early Triassic ammonoid faunal assemblages are present in the exotic blocks from Oman. Oman also has a rich diversity of rudist bivalve fossils. A site in the Ramlet Ghunaim in the Wilayat of Muhut is considered one of the best sites for these in the world. At least four different types of rudist shells can be identified in the Ramlet Ghunaim.

5 The Cenozoic Rocks and Life in Oman

The significant evolution and spread of mammals is one of the most prominent features of life of the Cenozoic Era. In Aydam, in the mountains of the southern part of the Governorate of Dhofar, four Omanis, out hiking, stumbled across this wonderful fossil of an elephant-like animal that lived in Dhofar about 35 million years ago.

M. H. Al Kindi, *Evolution of Land and Life in Oman: An 800 Million Year Story*,
https://doi.org/10.1007/978-3-319-60152-6_5

5.1 Earth During the Cenozoic

The Cenozoic Era spans a time period of about 66 million years, from the end of the Mesozoic up to now. It is often known as the age of mammals, because of their diversity, spread and growth in size during this period after being relatively limited in number and very small in size during the Mesozoic, when the dinosaurs were still living. The extinction of dinosaurs, which dominated the land during the Mesozoic, promoted the diversity of mammals and allowed them to occupy almost every terrestrial and marine environment. However, diversification was not limited to mammals. Birds, amphibians and smaller reptiles also changed and grew in size. In the shallow seas, new marine species took over, and finally—after 99.9% of the earth's present age—the most dominant creature ever found on earth appeared: Homo sapiens.

There are several different divisions for the time periods of the Cenozoic Era. The recently accepted one divides this era into three periods, known as the Palaeogene, the Neogene and the Quaternary. Previously, the Palaeogene and Neogene were known as the Tertiary; however, since 2004 this name is no longer recognized as a formal unit of time by the international community, although it is still used unofficially. The Palaeogene spans a time period of 43 million years, whereas the Neogene and Quaternary span 21 and 2.6 million years, respectively. These periods are further divided into five epochs—the Paleocene, Eocene and Oligocene of the Palaeogene, the Miocene and Pliocene of the Neogene and finally the Pleistocene and Holocene of the Quaternary (Fig. 5.1). Although it covers the shortest period of geological time, the Cenozoic has more detailed subdivisions largely due to our greater understanding of the climatic changes, tectonic events and evolutionary process that occurred during this era—the one we live in.

At the beginning of the Cenozoic, the continents were very similar to their present distribution. The African and South American plates moved farther apart as the Atlantic Ocean grew wider along its mid-oceanic ridge. The Arabian Plate separated from Africa along the spreading zone of the Red Sea, later colliding with the Eurasian Plate to form the Zagros Mountains in southern Iran, as well as the Makran Mountains and Al Hajar Mountains in Oman (Fig. 5.2). Both the Australian and Antarctic plates drifted away from Africa, with Australia moving north and Antarctica moving towards the South Pole. China collided with Asia and the Indian Subcontinent finally collided with Asia, after moving north for a long period. The Himalayas started to be uplifted at the collision zone. South America moved to the north away from the South Pole and connected with its neighbouring continent, North America. All the oceans in the southern hemisphere were connected right around the globe.

Since the beginning of the Cenozoic, the climate has been in a phase of steady and long-standing cooling and drying until today (Fig. 5.3). However, this cooling phase is often punctuated by short warmer periods during which the level of CO_2 increases in the atmosphere and the concentration of oxygen decreases.

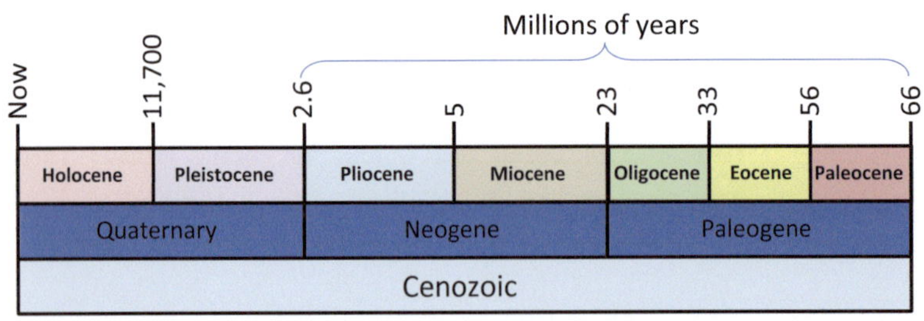

Fig. 5.1 Time periods and epochs of the Cenozoic

Fig. 5.2 Planet earth as it looks today, and the tectonic events that still occur around the Arabian Plate. They are directly related to the opening of the Red Sea at a rate of about 2 cm per year, the northward movement of Arabia and its subsequent collision with the Eurasian Plate along the collision zone in the Zagros Mountains in Iran. This collision zone was once the southern limit of the pre-existing Neo-tethys Ocean. The red line represents the collision zone between the two plates and the red arrows indicate the direction of movement. Arabia is also bounded by major fracture or fault zone from its east (Owen-Sheba Fault in the Indian Ocean) and west (Dead Sea Fault Zone). However, most of the tectonic forces and stresses that cause and result from the movement of Arabia are consumed within the boundaries or margins of the plate and almost all earthquakes occur along these margins

The most famous global warming event in the Cenozoic occurred during the Eocene epoch, when the climate reached its thermal maximum for unknown reasons. Another, smaller phase of warming occurred during the Oligocene. This phase is often attributed to the separation of South America from Antarctica, the formation of a mid-oceanic ridge between them and the partial melting of ice in Antarctica. Understanding the rise of global temperatures during such periods may be helpful in understanding the causes and effects of the earth's present phase of global warming.

From the Miocene until today, the earth has been continuously cooling. This resulted in the repeated glaciations periods of the Pleistocene, commonly known as ice ages. During these glaciation periods, ice covered great parts of the earth's poles, North America, South America, North Europe, Asia and almost all the high mountains around the globe. The ice ages contributed to the migration of animals across continents, but also caused small pulses of extinction. During the Pleistocene glacial events, sea levels frequently dropped by more than 100 m compared to current levels. The current

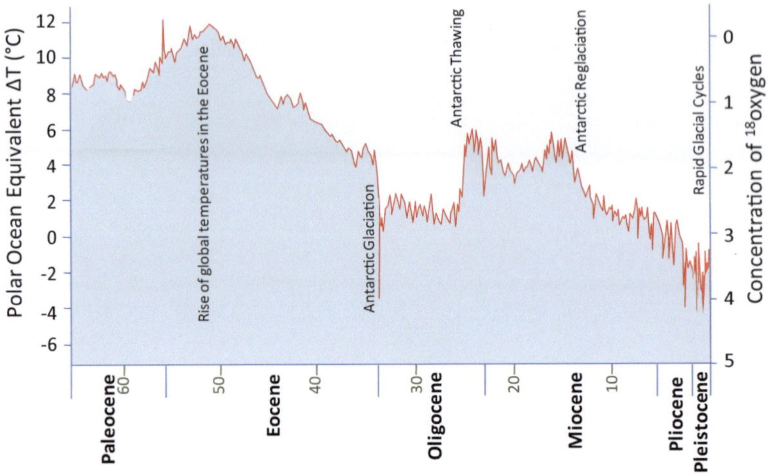

Fig. 5.3 Glaciation periods are usually determined from the percentage of oxygen isotopes in the composition of foraminifera fossils (Hansen et al. 2008), because the percentage of heavy oxygen isotopes (O18) increases and the concentration of light oxygen isotopes (O16) decreases when the temperature of seawater is lowered during glacial periods. There has been a general decrease in global temperatures throughout the Cenozoic with a relative temperature rise in the Eocene

instability of the earth's climate—especially since the last ice age—is causing worldwide concern. There has been an increase in global temperatures of almost 1 °C in the last 100 years or so, and there is the possibility of a further increase of about 2 °C during this century. There are only two large areas on earth today covered by ice, one in Antarctica and another around the North Pole.

Following the mass extinction at the end of the Mesozoic, the nearly vacant-of-life marine and land ecological niches supported the evolution of new life forms in the Cenozoic, making Cenozoic life substantially different from the Mesozoic. Many forms of life became abundant on land—vertebrate animals, such as mammals, birds and reptiles, invertebrate species, such as insects and gastropods, and plants, including flowering plants. The proliferation and diversity of animals and plants during this era were probably supported by the split of supercontinents into smaller continents, similar to their distribution today. For instance, marsupial animals like kangaroos developed in Australia and became endemic to its separate land mass, whereas other placental mammals formed the main type on many other continents.

In the early Cenozoic, many orders of the Mammalia class appeared. These included the Rodentia (e.g. mice and rats), Carnivora (e.g. cats and dogs), Cetartiodactyla (e.g. whales), Ungulata or hoofed mammals (e.g. horses and camels) and the Primates (e.g. monkeys). These mammals commonly lived around river deltas and coastal plains as evidenced by the abundance of their fossils in places like Al Fayoum in Egypt and the Oligocene rocks in the Dhofar Mountains. However, many of these early primitive species became extinct or developed into new forms during the Palaeogene, allowing new species to appear in the Neogene.

The new species became more dependent on grass instead of the tree branches that were prevalent at the beginning of the Cenozoic. This shaped the evolution of many groups, including mammals and birds. Most of the modern mammals gained their current characteristics during the beginning of the Neogene. The Cenozoic was also distinguished by the appearance of large birds, some larger than the average human. Snakes also became widely diversified and distributed in the Cenozoic.

Cenozoic marine life was also full of whales and different fish, as well as molluscs and protozoa. Corals were widespread along the tropical belt. However, the most distinguished organism that appeared in the Cenozoic is the human race, which emerged within the last two million years to become widely distributed across the planet by the end of the Pleistocene and the beginning of the Holocene, around 10,000 years ago. Since then, humans have massively contributed to the disappearance of many animals, including some large mammals (e.g. mammoths), particularly after developing new advanced tools and technologies for hunting.

5.2 Oman During the Cenozoic

After the emplacement of Ophiolite and the deposition of the Al Khawd Conglomerate, a shallow sea covered most of the present land mass of Oman and the Arabian Peninsula in the early Cenozoic, at a time when the earth was much hotter than today. Thick, successive layers of limestone were deposited in these shallow seas (Fig. 5.4). The limestone rocks of the Cenozoic can be seen across Oman and they usually have a light-yellow colour, as seen on the sides of Al Hajar Mountains in northern Oman, the desert plains of Central Oman and the western part of the Dhofar Mountains. These rock layers are often cut to construct new roads such as the Muscat Express Road, Qurayyat-Sur Road and Ash Shuwaymiyah-Sadah Road. The road cut to Al Khawd Village in As Seeb goes through the three main rock units that were

Fig. 5.4 Limestone layers from the Palaeogene epoch of the Cenozoic in the Wilayat of As Seeb. These layers are clearly visible along the new Express Road in Muscat and cover considerable parts of the Al Hajar Mountains, Dhofar Mountains and the desert plains of central and South Oman

Fig. 5.5 Volcanos from the Cenozoic in Al Ashkharah in the Wilayat of Jaalan Bani Bu Ali, where at least 10 different volcanoes erupted around 40 million years ago. Top: a volcano with eroded crest. Bottom right: black basalt at the top of a small volcano. Bottom left: pillow lava. Cenozoic basaltic extrusions are also present in the Wilayats of Al Duqm and Muhut in Central Oman

deposited in the Early Cenozoic. Cenozoic limestones are the most important water aquifers, particularly in Central Oman. Many famous caves (e.g. Majlis Al Jinn, Taiq Cave and Tiri Cave) and sink holes, particularly in Dhofar, formed because of the dissolution of these limestone rocks by fresh water. They contain a large variety of marine fossils, including various types of bivalve, coral and gastropods. They also contain significant quantities of gypsum and coal layers.

The deposition of limestone rocks continued in north and South Oman for most of the Palaeogene. During this epoch, the so-called Masirah Ophiolite was emplaced on a few areas of the eastern coastal shore of Oman, including Masirah and Ras Madrakah. A number of volcanoes formed along the same coast, in the Wilayats of Jaalan Bani Bu Hasan, Jaalan Bani Bu Ali and in the Al Huqf area in Al Wusta Governorate. These volcanoes erupted nearly forty million years ago, as black basalt originated

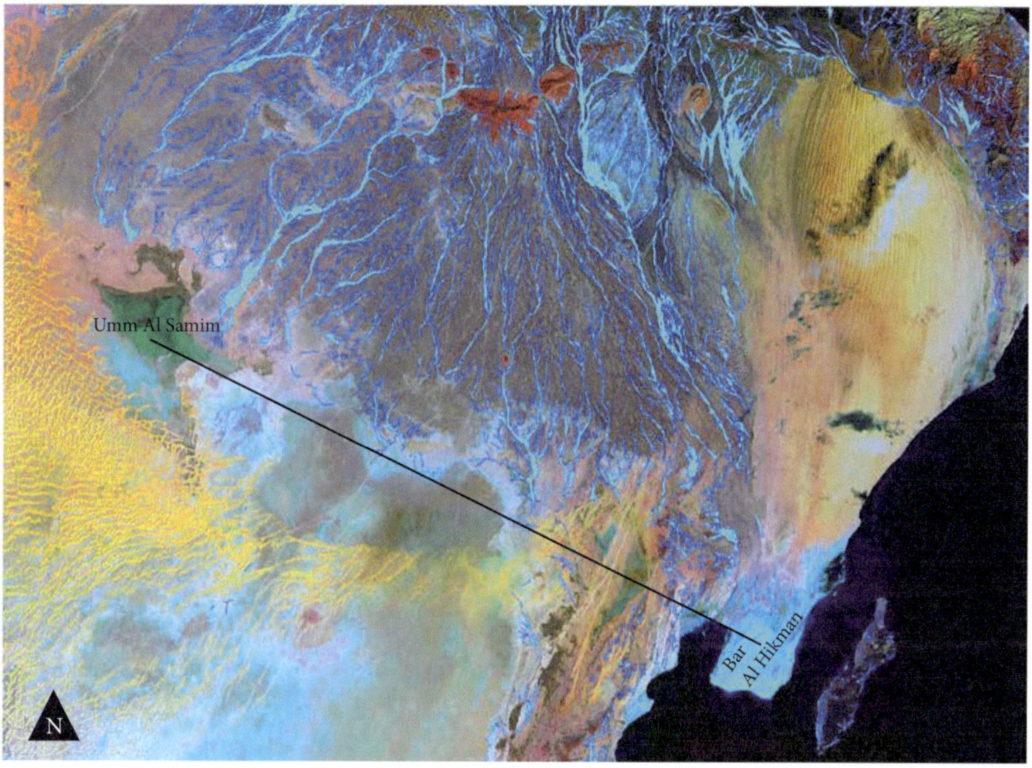

Fig. 5.6 Wadis of northern Oman run southward and dispose water along a depression zone that extends from Barr Al Hikman to Umm Al Samim. Many sand dunes terminate along this zone as well. The line between these two regions extends for about 300 km. The topography, recent sediment deposition, fossil content and ancient human artifacts around this area indicate that it was once rich in water and probably had permanent extensive lakes which are filled today by sabkha (Arabic word for a salt flat)

from great depths at the boundary between the crust and the mantle. It penetrated the overlying sedimentary rocks as it gushed to the surface (Fig. 5.5).

With the beginning of the Neogene Period, the sea retreated from most parts of Oman, except the southern part, and the deposition of limestone layers ceased. Subsequently, the exposed rocks of Ophiolite and sedimentary rocks were uplifted—probably as a result of the collision between Arabia and Eurasia—and they were weathered and eroded by rivers and wadis that re-deposited the eroded debris as sandstone or a conglomerate of pre-existing Ophiolite and sedimentary clasts cemented in a fine matrix. The ancient wadis of the Early Neogene were characteristically wide and straight. This geometry usually suggests continuous and abundantly flowing water. The deposition of limestone continued in South Oman during the Neogene.

The deposition of conglomerates and sandstone by rivers and wadis continued during the Quaternary Period (Fig. 5.6). The conglomerate clasts as seen on the edges of many wadi streams in Oman are often more than 20 cm long, indicating strong ancient wadi flows. There are also vast sabkha or playa areas along a zone extending from Bar Al Hikman to the sabkha of Umm

Fig. 5.7 Sabkha in Central Oman, on top the interaction of the Rub Al Khali sand dunes and the salt flat of Umm Al Samim Sabkha in the Wilayat of Ibri below extensive salt flat area in Bar Al Hikman in the Wilaya of Muhut in Central Oman

Al Samim, which represents the largest inland sabkha in Oman today. The ground elevation is just 59 metres above sea level. These sabkha represent inland depressions were wadis and underground flows discharge water. As the surface and percolated artesian water evaporates, polygonal salts formations occur. The layers of these salt pans may be several centimetres thick (Fig. 5.7). During rainy periods, the inland sabkha of Oman turned to large lakes around which

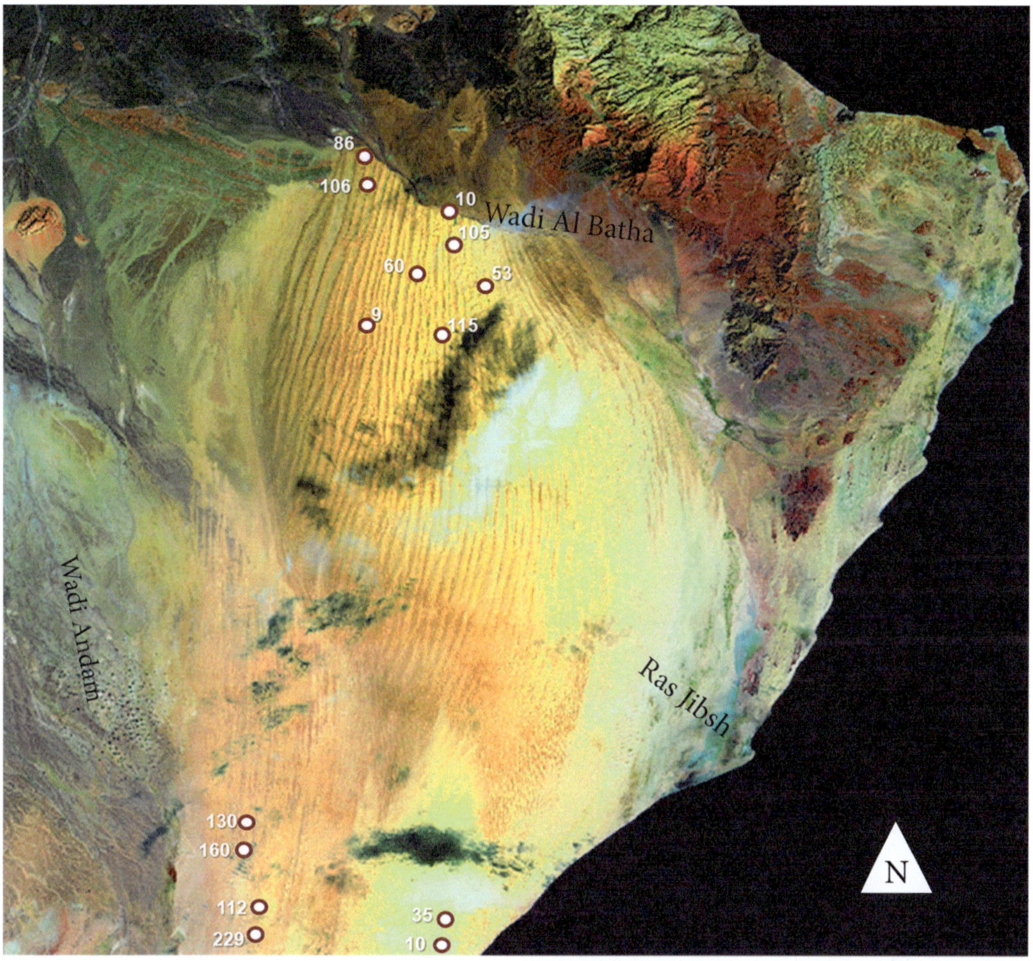

Fig. 5.8 Ash Sharqiyyah Sands from Space. The numbers represent the ages of the sand dunes in thousands of years as analysed by Prof. Ken Glennie using the OSL dating technology. The total length of the sands from north to south is about 150 km. The northern limit of the sand-dune mass is truncated by the Wadi Al Batha. The sand dunes probably have not extended significantly in the last 100 thousands years. The largest longitudinal dunes are present in the northern and central parts, where individual dunes may extend for about 100 km

humans lived—evident from ancient human artifacts.

The glacial periods of the Pleistocene epoch, representing the beginning of the Quaternary Period, were characterized by a dry climate with strong monsoon winds. Extensive sand dunes were deposited in different parts of Oman including the Ramlet Al Wihibah and Rimal Ash Sharqiyyah Sands, the sand dunes in Muhut, Bawshar dunes and some of the sand dunes on Al Batinah plain in northern Oman (Fig. 5.8). During the glacial periods of the Pleistocene, the global sea level dropped by at least 100 m and the coast regressed from the present shore by hundreds of metres. As a result, the strong winds eroded the exposed carbonate beds of the sea and re-deposited the carbonate fragments as huge sand dunes—often far inland and resembling

Fig. 5.9 Number of meteorites found in the desert plain of Oman, found by a group consisting of the National History Museum of Switzerland, University of Bern and the Directorate of Minerals in the Ministry of Commerce and Industry

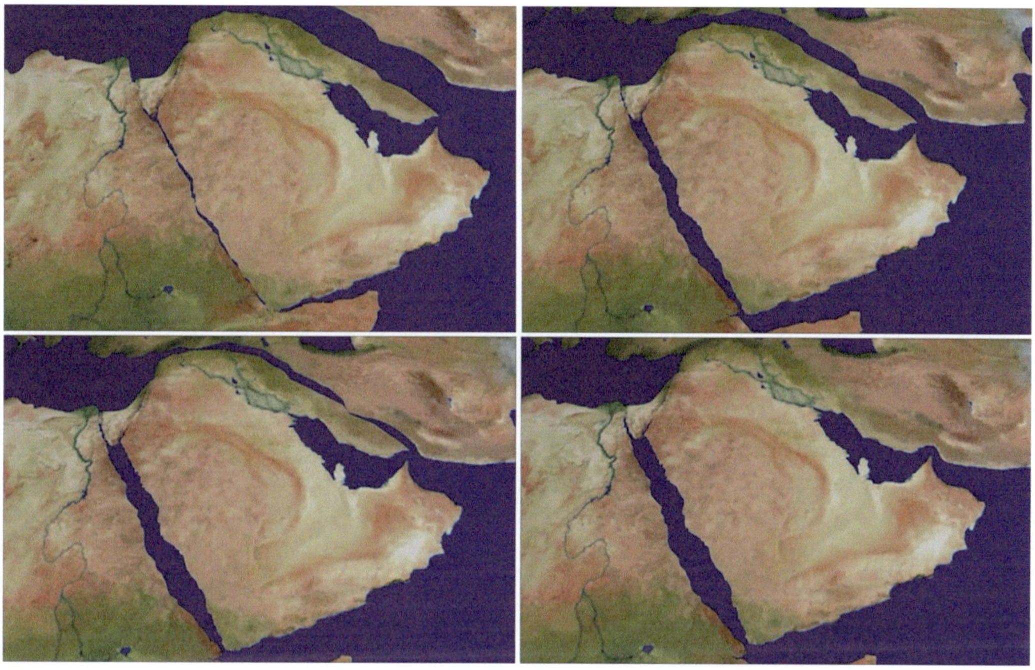

Fig. 5.10 Separation of Arabia from Africa since the opening of the Red Sea about 30 million years ago, and the subsequent collision between Arabia and Eurasia as a result of the northward movement of Arabia. The current spread rate of the Red Sea is about 2–2.5 cm

great mountain chains. The sand dunes in Bawshar, for instance, were most likely deposited during the glacial periods of the Pleistocene and their accumulation ceased about 10,000 years from present, during the last glacial period.

The Pleistocene Epoch ended with a glacial period—the last one in the Quaternary. During that time, the Tigris and Euphrates rivers flowed all the way to the Straits of Hormuz, near the Musandam Mountains, discharging water into the Sea of Oman. The continental area of the Arabian Gulf was not yet flooded with seawater. At its coldest extremity, the temperature along the Tropics was probably around 5–8 °C and was in Oman about 1–6 °C with very strong and persistent global winds. As sea levels rose during the end of the last glacial period, about 14,000 years ago, the sea gradually transgressed towards the land, at a rate of about four kilometres a year. In fact, it advanced as far as the ancient town of Ur, in Iraq, but then retreated to its present position.

During the Holocene Epoch—the time in which we live today—a number of meteorites from Mars, the moon and nearby orbiting asteroids fell around Oman. The clasts of these meteorites are preserved in many places in the desert plain of Central Oman. As the effects of rainfall and wadi flow are significantly reduced, these rocks are preserved for long periods without being moved by water flow or buried by new sediment deposition (Fig. 5.9). The desert plains of Central Oman were also unlikely to have been affected by major tectonics since the deposition of the limestone beds and so are considered one of the best places around the world for meteorite rocks. The nature of asteroids, Mars and the moon has been studied from samples found in Oman. So far, more than 5,000 pieces of meteorites, weighing several tons, have been found, making Oman number one for meteorite discoveries.

The Cenozoic Era witnessed significant movements in the final formation of Arabia. During the Oligocene, the Arabian Plate separated from Africa along the spreading zone of the Red Sea (Fig. 5.10). A rift was also created along the Gulf of Aden spreading line, causing the creation of the Dhofar mountain ranges. As the Arabian Plate continued to move north during the Miocene (Early Neogene), it collided with the Eurasian Plate pushing up the Zagros Mountains of southern Iran.

A subduction zone formed between the northern part of the Sea of Oman and Pakistan, now known as the Makran subduction zone. At the same time, the final uplift of the Al Hajar Mountains began, pushing these northern ranges to their present elevation.

Today, Oman—with the rest of Arabia—is still moving northeast at an estimated rate of about 2 cm per year, primarily because of the continued widening of the Red Sea at almost the same rate. As it heads north and pushes up against Eurasia, the northern edge of Arabia (the Musandam Mountains) is subsiding a few centimetres annually, while some northern and eastern parts of the coastal margin of Oman such as Al Batinah, Muscat and Sur are rising by about 2 cm each year.

5.3 Organisms Living in Oman During the Cenozoic

With the onset of the Cenozoic, life was substantially different from now, following the mass extinction of the Cretaceous and the survival of selected species from each organism group. It is not particularly understood why certain species went extinct while others managed to survive. In general, earth was dominated by small fauna at the very beginning of the Cenozoic. However, life has continued to evolve and diversify since then. Different types of mammals, insects, birds, fish, coral, molluscs and plants appeared. Following the wide distribution of continents throughout the planet, various organisms appeared in different parts. It is hard to document all the known types of organisms that lived or continue to live in Oman in the Cenozoic. This section records the most spectacular known Cenozoic fossils in Oman.

Mammals appeared on earth during the Triassic (the first-time period of the Mesozoic); however, their size probably failed to exceed the size of a big domestic cat and there were only nine defined orders. After the split of continents and the extinction of dinosaurs during the Mesozoic, dry land was ripe to host more complex and diverse mammals. During the Cenozoic, known orders of mammals reached 30—comprised of almost 5,500 species. Some species were several metres in length, like whales and some types of ungulates. All life on earth was subject to the vagaries of the fluctuating climatic and environmental conditions of the Cenozoic. Many species were wiped out in extremes of glaciation while others diversified in tropical climates. Mammals were at their most varied in the period between the beginning of the Eocene and the middle part of the Miocene.

Although the fossil record of mammals is confined to teeth, skulls and bones, it is sufficient to determine the type of the mammal and its size. Many species of mammals, from various periods, have been found in the Cenozoic rocks of Oman. Of those over 10,000 years old—prior to the Holocene—are included Proboscidea, animals with long, mobile noses like elephants, a horned animal, Arsinoitherium, akin to a rhinoceros, and various Primates including the first humans.

5.3.1 Proboscideans

The proboscideans, named after their long nose or proboscis, first appeared about 60 million years ago at the start of the Cenozoic Era. They were less than a metre tall and probably had no

Fig. 5.11 Schematic drawing of the Omanitherium dhofarensis or the Barytherium Omansi, with few details of the upper and lower jaws

trunk. However, over the various Cenozoic epochs, they diversified and dominated large parts of the planet. They reached more than 4 m tall and over 10 m long and developed large tusks, muscular trunks and thick, column-like legs. The Proboscidea order includes many families, but only one family, Elephantidae, is still living today. This family includes two living genera, the African Elephant (the largest land animal in the world today) and the Asian Elephant. Different types of proboscideans became extinct throughout the Cenozoic. The most recent were mammoths and mastodons, which became extinct in the last glacial period, about 10,000 years ago, probably as a result of the over-hunting by humans.

Among the early proboscideans were Barytherium and Deinotherium. Remains of these animals were found in south and Central Oman. The fossil distribution of these animals in Africa, Asia and Europe indicates that Oman formed a bridge for many types of proboscideans during their journey from Africa to Eurasia.

5.3.1.1 Barytheriidae

In Aydam, in the mountains of the southern part of the Governorate of Dhofar, four Omanis, out hiking, stumbled across a wonderful fossil. It was estimated at over 35 million years old, dating back to the Oligocene epoch of the Palaeogene, and it belonged to an extinct type of proboscidean—an order containing mammoths, mastodons and elephants. Only the elephant family survives to the present day, with just two species—the Indian and African elephants. Early proboscideans probably lived on the coastal fringe of shallow seas, similar to the many khors and lagoons along the coast of Oman today. They

Fig. 5.12 Fossils jawbone and teeth of Omanitherium dhofarensis or Barytherium Omansi. The jawbone on top is about 30 cm long and the incisor on the bottom left is about 5 cm long

first appeared during the Paleocene and became highly varied during the Eocene and Oligocene. The fossils found in Dhofar most likely belong to the Barytheriidae family—the first large, primitive type of proboscideans to appear on earth (Fig. 5.11).

Many fossils of the Barytherium proboscideans have been found in Al Fayoum in Egypt; however, the Omani fossils display different characteristics when compared to all the other members of the Barytheriidae family discovered to date. Among the differences is the presence of two largely spaced incisors (Figs. 5.11 and 5.12). Such anomalies have led to a separate classification of the Omani Barytherium into a new species of proboscideans. The new species found in Dhofar is named Omanitherium dhofarensis or Barytherium Omansi. The suffix therium means 'wild animal' in Greek.

Fig. 5.13 Reconstruction of Deinotherium compared to an average-tall person. Some fossils of this animal indicate even larger bodies and longer tusks. It may have been the third largest animal that lived on earth

Fig. 5.14 Fossils of Deinotherium teeth in the Al Ghabah area of Central Oman (found by Martin Pickford). They are about 5 cm long

Fig. 5.15 Artist's impression of a Gomphotherium

5.3.1.2 Deinotherium

On the eastern side of Al Ghabah in Central Oman, at the southern limit of Al Dakhiliyah Governorate, other proboscidean fossils have been found. They include fossils of a Miocene proboscidean, some 17 million years old, of the genus Deinotherium. Deinos means 'terrible' in Greek. These creatures lived on earth for about 14 million years, becoming extinct during the early Quaternary (about 2 million years ago). Their fossils continue to spark debate because of the downward-curved lower tusks (Fig. 5.13). It is not clear how deinotheriums used these tusks. They might have used them to dig up shrubs from the ground, to drag tree branches to their mouths, as a means of defence or as a tool for dragging their bodies out of water. The tusks of deinotheres are also different because they emerge from the lower jaw. Deinotheres had a relatively small, flat-topped skull. Their teeth (Fig. 5.14) were adapted to shear and grind tree branches. Deinotheres lived in Africa, Asia and Europe. Native species of deinotheres appeared on various continents, such as Deinotherium Giganteum in Europe, Deinotherium Indicum in Asia and Deinotherium Bozasi in Africa.

The fossils of Deinotherium in Oman include many complete teeth, fragments of bones and poorly preserved tusks that were mainly found in the Al Ghabah area. From a biogeographic sense, this area is highly important because it is the closest site we know to Eurasia with this fauna. Deinotheriums living in Al Ghaba were probably the first to cross from Africa to Eurasia 20 million years ago.

5.3.1.3 Gomphotherium

Other proboscideans also lived in Oman. Among these is the famous animal Gomphotherium which appeared on earth about 14 million years ago during the Miocene and became extinct after more than eleven million years during the Pliocene about 3.5 million years ago (Fig. 5.15). Gomphos means 'welded' or 'nailed' in Greek. Gomphotheriums were very similar to modern-day elephants but had four tusks—two on the upper jaw and two on the lower. The upper tusks pointed gently to the body, whereas

Fig. 5.16 Top: a number of gomphotherium teeth found in Al Ghabah by a group of paleontologists from the Natural History Museum of Paris. The samples are currently preserved in the Natural History Museum of Oman and each sample is between 7 and 12 cm long. Bottom: New samples of gomphotherium teeth found recently in the Al Ghabah area. The area remains very attractive for further discoveries of Cenozoic proboscideans

Fig. 5.17 Left: a proboscidean tusk from the Cenozoic rocks of Al Ghabah. The tusk is preserved and displayed in the Natural History of Oman and it is about 10 cm long. Right: a few marine fossils in Al Ghabah. They include a fossil fish vertebra top left and fragments of oysters. Along with fish bones and mammal fossils, the area also includes a lot of crocodile and turtle fossils. Overall, the Cenozoic rocks of Al Ghabah represent alternation between marine and terrestrial deposits and suggest that the area was once very close to the beach and was probably covered with mangrove swamps. It served as a bridge for mammal migration in the early Miocene, about 20 million years ago

Fig. 5.18 Artist's impression of a arsinoitherium. This animal lived during the late Eocene and early Oligocene. Its fossils in Oman are probably the oldest known in the world, as they date back to more than 35 million years ago

the lower ones were upwardly shaped and were probably used as shovels to dig up aquatic vegetation near lakes or swamps. Their skulls were also more elongated and flatter compared to elephants today. They fed on trees and shrubs or mangroves living on the shores of lagoons or coastal inlets.

Gomphotheriums also had elongated and low skulls and probably long trunks to allow them to pick their food (Fig. 5.15). However, in general, they were smaller in size than the giant proboscideans of the Miocene. The evolution history of gomphotheriums is still not well understood. They are considered part of a second radiation of proboscideans after the earlier ancestral group (e.g. Arytherium and Deinotherium). Gomphotheriums were able to spread across a vast area of the planet, including North America, Europe, Africa and different parts of Asia, having the ability to live in various climates and utilize narrow land bridges to migrate from one continent to another. They reached their peak diversity about 10 million years ago before reducing in number at the end of the Miocene.

The fossils of gomphotheriums in Al Ghabah include many complete teeth (Fig. 5.16), fragments of bones and poorly preserved tusks. Al Ghaba was probably then a wetland, along the shoreline of a shallow sea.

5.3.1.4 Other Fossils of Elephants and Marine Species in Al Ghabah

The fossils in Al Ghabah provide valuable information about the evolution and way of life of proboscideans during the Miocene, 15 million

Fig. 5.19 On top a lateral view of an Arsinoitherium fossil found in the Dhofar Mountains, about 30 cm long. It is preserved at the department of Earth Sciences at Sultan Qaboos University. Below are two teeth of possibly Arsinoitherium, found by Martin Pickford, each of about 4 cm long. These samples are preserved in the Natural History Museum of Oman

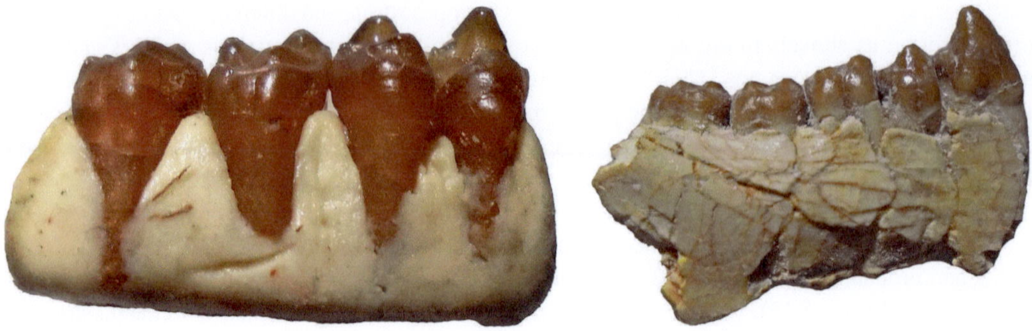

Fig. 5.20 Primate fossil teeth from the Cenozoic Era found in the Dhofar Mountains by a group of scientists from the Natural History Museum of Paris. Each sample is about 5 cm long

years ago, and there are still great opportunities to find excellent fossils in this area. Along with proboscideans, the Al Ghabah area also contains the fossils of crocodiles, turtles and fish (Fig. 5.17). It seems that this area had a tropical climate during the Miocene, with many lakes and grasslands, in stark contrast to its current desert climate. The rocks of these lakes are very bright in colour and contain numerous mammal fossils. Unfortunately, they are rarely well preserved because of the gypsum growth and crystallization that usually destroys any remains.

It is believed that Oman, as evident from the Miocene rocks of Al Ghabah, formed a bridge for many mammals, particularly proboscideans, during their journey from Africa to Eurasia and Asia, in a major dispersal event known globally as the 'Proboscidean Datum Event', which occurred in the early Miocene, about 20 million years ago. This event was first identified from the abrupt occurrence of proboscideans (e.g. Deinotherium and Gomphotherium) and other mammals in the continental layers of Europe and Asia.

The proboscidean fossil collections in Oman also include fragments of mastodons, which were very similar to mammoths, but had shorter legs, stronger muscles and longer bodies. They first appeared at the end of the Miocene and lived until the end of the Pleistocene 10,000 years ago, during the last ice age. The reason of extinction is not precisely known, but it could well be related to the expansion of humans, particularly on the American continents, where stone tools and mastodon bones have been found together.

5.3.2 Horned Animals

Among the fossil remains of the Dhofar Mountains, explorers discovered a bone belonging to a creature called Arsinoitherium (Fig. 5.18). These were herbivores, similar to a modern-day rhinoceros as they bore a massive pair of horns above the nose. However, they had thick, columnar legs and five-toed feet which are closer to the modern-day elephant.

A complete skeleton of Arsinoitherium was found in Al Fayoum in Egypt, named after Queen Arsinoe who ruled the area in Ptolemaic times. However, the fossil found in Oman, from Arsinoitherium zitteli, might actually be the oldest fossil evidence of the genus as it dates back 40 million years (Fig. 5.19). These animals probably lived in similar environments to Omanitherium dhofarensis among mangrove swamps and coastal lagoons in Dhofar. They were able to travel easily between North Africa and Arabia as the Red Sea did not exist then. They completely disappeared during the Early Oligocene, about 25 million years ago.

Fig. 5.21 Footprints made by an artiodactyl ungulate about 7 million years ago (Middle or Late Miocene) in a coastal cave in Musandam. The trackway is 6.5 m long and it includes 6 impressions in the ceiling of the Al Jissah Cave in the Wilayat of Daba (first reported by Schulp et al. 2011). The animal has two hooves and the size of individual footprints (well over 25 cm in length) indicates that the animal was a large artiodactyl, possibly a primitive giraffe (e.g. Canthumeryx). The white rocks that make up the cave are aeolian sands, most likely rich in white carbonate grains, that have been most likely deflated from the previously exposed platform of the shallow sea. Erosion has removed the underlying rocks of the cave ceiling and exposed the negative impressions of the footprints. Similar but less preserved trackways of ungulates and proboscideans are present in Al Ghabah in Central Oman

5.3.3 Primates

Fossils of bones and teeth of early primates have been found near the Wilayat of Taqa and Aydam in the Governorate of Dhofar (Fig. 5.20). The oldest of these belongs to the Oligocene Epoch. Most of these fossil species were previously unknown and have been named after Oman or the area in which they were found, Omanodon and Shizarodon dhofarensis. These species became extinct about 10 million years ago, during the Miocene. Overall, there are about 12 different species that belong to 5 families of primates found in Dhofar. About 300 different teeth have been recovered so far.

5.3.4 Ungulates

Among the mammal fossils found in south, central and northern Oman are a number of ungulate fossils that belong to various epochs of

Fig. 5.22 Fossils of bones and a tooth (bottom right) of Canthumeryx found in the Al Ghabah by a group of scientists from the Natural History Museum of Paris. The samples on top are about 5 cm long

the Cenozoic. The ungulates represent a group of hoofed or toenailed animals that first appeared during the beginning of the Paleocene. By the Eocene, they became highly diversified and widespread on different continents. Examples of extant ungulates include horses, camels and giraffes. Most living ungulates are classified into two main types, Perissodactyla (odd-toed ungulates) and Artiodactyla (even-toed ungulates).

In a coastal cave in the Wilayat of Daba in the Musandam Governorate, a number of trace fossils that represent the footprints of ungulates (an artiodactyl trackway)—most likely a giraffe-like animal—were found (Fig. 5.21). It is not precisely known when this animal lived in Musandam, but evidence suggests that the fossils date back to the Middle or Late Miocene, about 7 million years ago. The trackway is preserved in aeolian sands that were deflated by the wind from an exposed platform of the shallow sea to the east. Similar trace fossils of ungulates and proboscideans can be found in the Al Ghabah area.

The recent surveys in the Al Ghabah area also revealed fossil remains of Canthumeryx (Fig. 5.22), which most likely represents an extinct type of giraffe, although no one is sure quite what it looked like (Fig. 5.23). It was one of hundreds of fossil bones, many of which have not yet been identified and fuel the efforts of explorers to make new discoveries about extinct mammals which once roamed the mountains and plains of Oman.

Among the ungulate fossils found in the Miocene rocks of Al Ghabah are the fossils of anthracotheriums, which resemble pigs or hippopotamuses, but with a narrower head and smaller body (Figs. 5.24 and 5.25).

Fig. 5.23 An artist's impression of a primitive giraffid artiodactyl known as Canthumeryx. Bone and tooth fossils of Canthumeryx were found in the Al Ghabah area in Central Oman. The Miocene trackway in the Al Jissah Cave of Daba, as shown in Fig. 5.21, was possibly made by a Canthumeryx

Anthracotherium means 'coal beast' since the first fossils were recorded in coal beds. They had wide feet, enabling them to thrive in muddy conditions around the edges of lakes and swamps. They first appeared during the Eocene and became extinct in the Miocene or Pliocene. They existed in Europe, Asia, North America and Africa. The marine and terrestrial deposits of the Cenozoic indicate a diversified fauna and flora far wider than mammals. Towards the end of this era, a number of ice ages were featured. Subsequently, animals like mammoths and mastodons evolved to cope with this climate. Mammals were generally larger than their present-day descendants. Early humans possibly migrated from Africa about 200,000 years ago.

5.3.5 Marine Animals and Tree Fossils in the Cenozoic

In Oman, the process of evolution has continued in seas and oceans, resulting in the appearance of many organisms. This is reflected in the Cenozoic rocks of Oman, which are littered with the fossilized bones and teeth of fish, crabs, gastropods, bivalves and so on (Figs. 5.26 and 5.27). This diversity could be related to an improvement in environmental conditions, such as increased nutritions. Moreover, the travertine and tufa deposits of the Cenozoic include, in various places of Oman, a wide range of fossil plants (Fig. 5.28).

Fig. 5.24 An artist's impression of an anthracotherium, which represents an extinct artiodactyl ungulate mammals that lived from the Eocene to probably the late Miocene. Anthracotherium is often considered as an ancestral form of hippopotamus. These two animals are in turn possible ancestors of whales

Fig. 5.25 Fossils of anthracotheriums. They include two bones, a number of teeth and a possible coprolite (bottom right). The white bone on the left is about 5 cm long

Fig. 5.26 Simple representation of the evolution of trees and marine animals during the Cenozoic. Whales first appeared about 50 million years ago. Dolphins emerged later in the Miocene. Flowering plants became the dominant flora on land, representing at least 80% of all plants. Almost all modern seed plants evolved in the Miocene. In Oman, various species of palm trees became dominant

5.3.6 Other Mammals (e.g. Hyraxes and Bats)

The Early Cenozoic rocks of the Wilayat of Taqah in Dhofar consist of many mammal fossils, many of which are not yet fully studied or properly identified (Fig. 5.29). They include fossils of hyraxes (order Hyracoidea), which first appeared in the Eocene and continue to live in the mountains of Dhofar in large numbers. They also include fossils of bats (order Chiroptera), which also appeared during the Eocene Epoch and found an ideal habitat in the large limestone caves of the Dhofar Mountains. Dhofar Governorate has already yielded many fossils of endemic mammals from the Early Cenozoic (particularly the Oligocene Epoch) and still has huge paleontological potential.

5.3.7 Summary of Life in Oman During the Cenozoic

The Cenozoic scenario is based mostly on three famous Omani fossil sites: Taqah and Thaytiniti (Dhofar area; Earliest Oligocene) on the left side of the landscape, and Ghabah (Early–Middle Miocene), on the right (Fig. 5.30). The drawing is made by Oscar Sanisidro.

Oligocene fauna is characterized by the embrithopoda cf. Arsinoitherium, an undescribed proboscidean similar to a Barytherium, the hyracoid Thyrohyrax meyeri, and primate Simiiformes (Oligopithecus spp., Catopithecus) and adapiforms of the species Shizarodon dhofarensis. Micromammals are represented by an undetermined Anomaluridae flying squirrel. The flora of the Oligocene scenario is based on the Al

Fig. 5.27 Various Cenozoic marine fossils. Top: coral fossils from the Wilayat of Qurayyat. Mid-left: a 35-million-year-old crab fossil, found by Al Khattab Al Mahrooqi near the Wilayat of Sur. Mid-right: a fossil gastropod from the Wilayat of Muhadha. Bottom: fossilized shark teeth from Ash Sharqiyyah South Governorate. The large size of many Cenozoic fossils reflects the warm tropical seas they lived in

Fig. 5.28 Tree fossils from the Cenozoic Era in various parts of Oman. Top: a fossil leaf of a palm tree in the travertine Cenozoic rocks of Fanja, similar to those that grow in many wadis in Oman today (top right). Middle: fossils of leaves from Fanja. Bottom: fossil palm-tree leaf (on the left) and other fossil leaves (on the right) from Al Jabal Al Akhdar

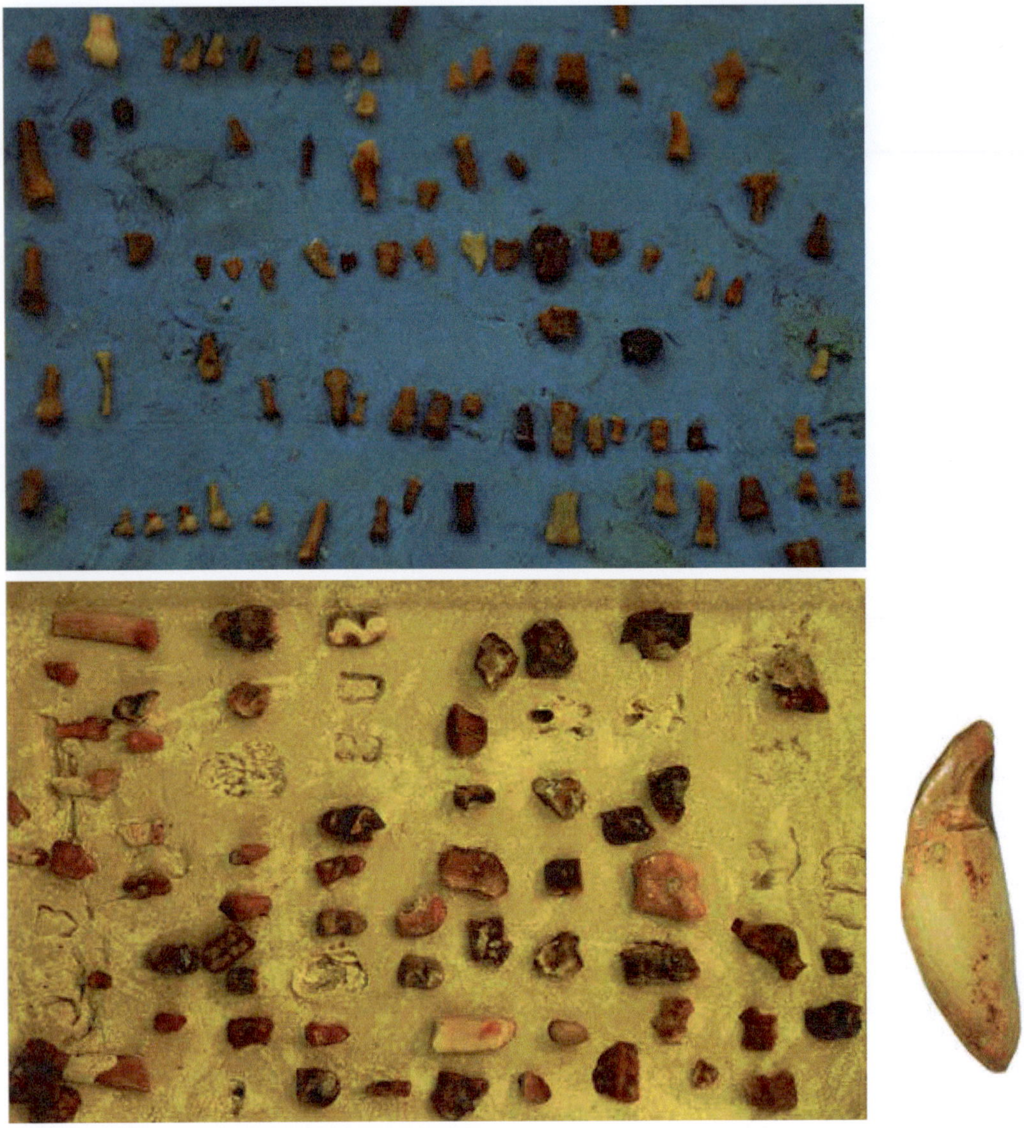

Fig. 5.29 Cenozoic mammals fossils from the rocks of the Wilayat of Taqah in Dhofar. On the top plate are fossils of hyracoid teeth and bones, and below are fossils of bat bones and other mammals. The samples are generally less than 1 cm long. The tooth on the right is for a Cenozoic carnivore (about 4 cm long). These fossils are found by a group of geologists from the Natural History Museum of Paris

Fayoum fossil site, being dominated by herbaceous grass and reeds of the genus Typha, palm trees of the species Palmoxylon aschersoni, and broad-leaf trees in the foreground and some leguminous trees (Leguminoxylon spp.) in the background.

Miocene mammals are represented by the small anthracotherid Afromeryx zelteni, the primitive giraffid Canthumeryx sp., the small elephantid Prodeinotherium sp., crocodiles (Tosmistoma-like), trionychid tortoises (cf. Cycloderma) and giant turtles of the genus cf. Geochelone. On the

Fig. 5.30 Summary of life during the Cenozoic as represented by the fauna and flora in Dhofar (to the left) and in Al Ghabah (to the right). The drawing is made by Oscar Sanisidro for this book

other hand, the Miocene flora depicted on the right side of the drawing is also dominated by dry herbaceous plants. In the background, there are some sparse Sonneratia-dominated mangroves. Lotus fossils (Gen. Nelumbo) have been found also in Al Fayoum, being represented associated with Miocene marshes together with reeds (Typha-like).

6 Humans

The arrival of humans on the planet changed complexity of life that dominated earth for millions of years. This picture is of one of the oldest graves in Oman. It dates back to the end of the 5th millennium BCE (about 6,000 years ago). It was found in Ras Al Khabbah on the eastern coast of Oman (Photo: Joint Al Had Project, Ministry of Heritage and Culture).

M. H. Al Kindi, *Evolution of Land and Life in Oman: An 800 Million Year Story*,
https://doi.org/10.1007/978-3-319-60152-6_6

6.1 The Emergence of Humans

After more than 99.9% of earth's present history, conditions were ripe for the first life form on the planet capable of performing complex thinking, speaking structured languages and developing sophisticated tools. Humans were clearly a step forward in the evolution of life on earth. As a species, they have witnessed and lived an infinitesimal portion of earth's history—mere seconds on the earth's clock (Fig. 6.1).

Despite the short existence of humans on earth, the order and complexity of life that had existed for millennia changed forever with their arrival. Here were creatures who could walk fully upright and had hands with an opposing thumb and flexible fingers providing them with a unique grip and the ability to perform intricate tasks with precision. Here were creatures with a finely structured larynx and delicate vocal chords capable of producing an array of sounds to express ideas and emotions and the ability to communicate in an array of complicated languages. Here were creatures with incredibly complex brains enabling them to learn, develop and form social relationships far in advance of than any other form of life on earth. But from where did they come?

The oldest fossils date to the Palaeocene, but there is a lack of conclusive evidence for when the evolution to the Homo genus occurred (Fig. 6.2). Fossils discovered so far are not only fragmentary and scarce, but also incomplete—often consisting of single bones or isolated teeth, making them unhelpful in positively deducing any behavioural patterns or lifestyles indicating the advent of modern man. On finding fossils belonging to antecedent Simiiformes, the transition to hominids was assumed via subsequent Hominini, although the topic was much debated and invited many interpretations because decisive scientific evidence was absent. Hominidae were depicted as Hominini, and early humans were drawn as Simiiformes in terms of shape and behaviour, with the aim of creating a mixture between the two species to persuade the audience that such transition had really occurred at some point. The actual fossils discovered so far may

Fig. 6.1 After completing more than 99.9% of earth's history to date, the conditions were ripe for the first organism capable of performing complex thinking, speaking structured languages and developing sophisticated tools. Humans were clearly a step forward in the emergence of life on earth. As a species, they have witnessed and lived an infinitesimal portion of earth's history—mere seconds on the earth's clock

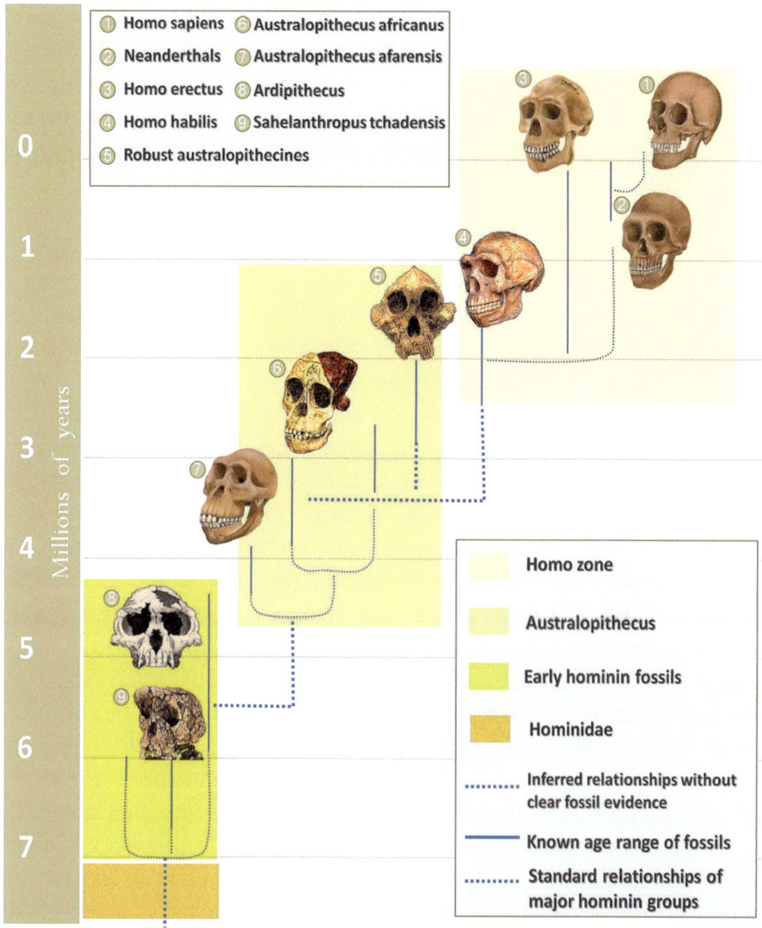

Fig. 6.2 Evolution of humans (redrawn and modified from a previous drawing of Jonathan Jones). A step in evolution occurred between the hominin and the human groups about two million years ago

actually be drawn and interpreted on the behavioural level in numerous ways. Meanwhile, ongoing scientific research is constantly revealing new information about fossils unearthed decades back. Unquestionably, we still know very little about our own origins and our ancient fossil record, and are still influenced by old presumptions made by early scientists over the last two centuries. We still need dedicated scholars and more in-depth scientific research in order to figure out what really happened and when.

Oman was not isolated from this human story or from the migrations made by man from one place to another since his emergence. There is a wealth of recent discoveries and scientific analysis to confirm the existence of humankind in Oman throughout all the different eras of history. However, just as with the fossils from all the geological ages, the majority of human fossils and the historical tools associated with them are still waiting to be discovered. We should redouble our efforts to find the evidence we need to piece together the story of life in Oman in a more complete way.

6.2 Classification of Humans

Humans have ranked themselves in the primate order of the Mammalia class because of their superiority over other organisms. Their bodies are more refined, their features better developed, and their brains more able to perform advanced mental tasks. While the primate order also includes monkeys, apes and lemurs, they are of a different branch from humans and have decidedly different capabilities. They also differ in their utilization of natural resources, as well as in the way they interact among themselves and with others.

Within the order is the Hominoidae family, which, besides man, includes chimpanzees, gorillas, orangutans and gibbons (Fig. 6.2). Around 18 million years ago, the line split, with the Hylobatidae (gibbon) diverging from the Hominidae line; 14 million years ago, supported by the earliest fossils of the line, there was another split, with the Ponginae (orangutans) diverging from the Hominini line. Around 7.2 million years ago, the line diverged again with Gorillini (gorillas) on one side and Hominini (the common ancestor of chimpanzees and humans) on the other. Some 4.7 million years ago, the final split between the Pan (chimpanzees) and Homo (human) lines is thought to have occurred, based on DNA and fossil evidence.

Two of the most famous Hominini fossils are named Ardi and Lucy. They belong to the genera of Ardipithecus and Australopithecus, respectively, and date back to 4.4 and 3.2 million years ago. The relationship between these two fossils and their counterparts has been under intense scientific debate since their discovery in the 1970s and 1990s; however, both display the fundamental characteristics of the hominin line: bipedalism, or the ability to walk upright on two feet.

The study of humans, past and present, is part of anthropology. However, the science concerned with past climate changes, the environment in which early humans evolved and the fossil record of our human ancestors is geology (Fig. 6.3). All humans today are part of one species, known scientifically as Homo sapiens. In Latin, sapiens means 'wise' and homo means 'human'. This nomenclature was given in the eighteenth century to distinguish the ambassador of modern humans from all other human species, yet when the name was first used, science was in its infancy and the mental capacity and behavioural patterns of extinct early humans were unknown. Even now, they can rarely be accurately deduced from the fossils found so far. We are the only extant species of the Hominini tribe, but from the beginning our ancient relatives sought to occupy every corner of the globe, migrating, developing, interbreeding and evolving into the supreme member of the primate order, H.sapiens.

The first of the Homo genus was Homo habilis, then Homo erectus and then Homo Heidelbergiensis which used to be called 'archaic Homo sapiens' and from whom the branches of Homo Neanderthalis and Denisova hominins derived. Finally came modern humans, sometimes designated Homo sapiens, to avoid confusion with earlier forms. However, many of the fossils of the Homo genus show variations, and it is difficult to tell whether a fossil belongs to one species with a familial mutation, or to another. The story is not one of direct linear descent, nor is the fossil record complete enough to draw our ancestral tree with complete certainty.

H.habilis appeared about 2.3 million years ago at the beginning of the Pleistocene epoch and disappeared about 1.4 million years ago. Since the discovery of the first H.habilis fossils in the middle of the last century, there has been controversy as to whether or not such a creature belongs to the Homo genus or to Australopithecus, particularly since there are a number of characteristics that distinguish it from modern humans—namely the long arms, small brain and short stature. Some have suggested renaming it Australopithecus habilis instead of H.habilis and even removing it from the human evolutionary series to become a separate and stand-alone branch.

H.habilis, whose name means 'handy man', was thought to be the first creature on earth who managed to utilize stone tools, many of which have been found in association with H.habilis fossils in Africa. They were probably used to hunt and butcher small animals. However, knapped stone tools have been found with Pliocene hominin fossils 3.3 million years old in Kenya (designated Lomekwi 3 knappers), reopening the debate as to where H.habilis fits into the human family tree. Certainly, H.habilis

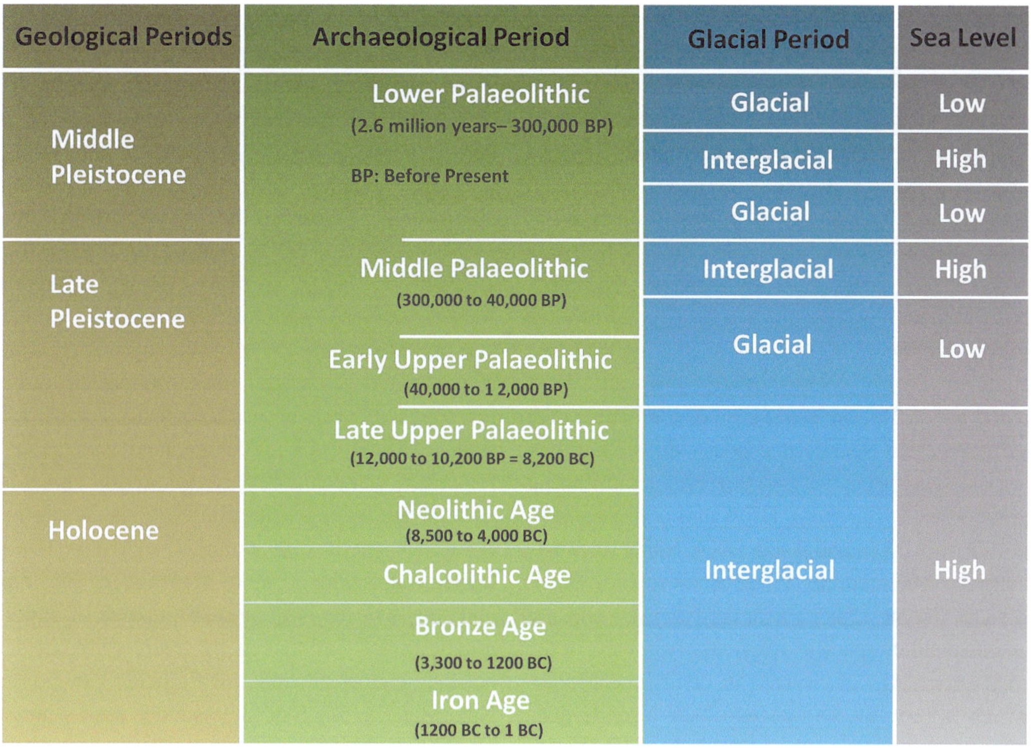

Fig. 6.3 Chronological correlation between geological and archaeological periods in comparison with the glacial/interglacial periods and the sea level variation

differed from apes in that he migrated from a forest habitat to live on open grassland. His brain was about the same size as a chimpanzee, but it has not yet been confirmed whether H.habilis was able to speak, but it is possible he developed a primitive language to communicate with the other members of his species.

Homo erectus, whose name means 'upright man', appeared about 1.9 million years ago and disappeared from the fossil record some 150,000 years ago or less. This was the first species to venture out of the continent of Africa, home of ancient man, to spread out across the earth. In addition to Africa, H.erectus fossils have been found in India, Indonesia, Georgia, China, Sri Lanka and Vietnam. The stone tools used by H.erectus are found all over the world and particularly in the south of the Arabian Peninsula. Fossils discovered in Georgia, located between Asia and Europe, are considered the oldest and the most significant because they are complete and date back 1.8 million years. In addition, the H.erectus skulls discovered in Georgia are so diverse in terms of shape and size that some scholars have suggested that the aforementioned H.habilis actually belongs to H. erectus.

H.erectus appears to have left Africa during the pluvial periods which turned the passage between Africa and Asia into pastures abundant with animals and water between 1.3 and 1.8 million years ago. This made it easier for H. erectus to cover long distances. Some other hypotheses, however, suggest that H.erectus emerged first in Asia and then moved to Africa. H.erectus was apparently able to make relatively elaborate stone tools, and it is not beyond the bounds of possibility that he was able to make primitive rafts used for short-distance sailing. Some geological evidence from mud sediments indicates—albeit inconclusively—that H. erectus used fire over one million years ago and developed a primitive language within small nomadic hunter-gatherer communities.

The fossils of H.neanderthalensis—Neanderthals—were first discovered in the Neander Valley in Germany. These early humans spread between Western Europe and the northern and central parts of Asia. They probably emerged at least 250,000 years ago and died out some 40,000 years ago. They were possibly out-competed by H.sapiens who first arrived in Europe around 45,000 years ago, sharing the same continent with Neanderthals for at least 5,000 years. Recent genetic evidence from DNA extracted from Neanderthal bones shows that they not only coexisted, but also interbred, contributing around 2% to European DNA and a smaller proportion to African and Asian DNA. Some scientists consider Neanderthals to be a subspecies of H. sapiens (Homo sapiens neanderthalensis) and not an independent species—particularly in view of the rock engravings discovered in 2012 in Gibraltar, which appear to demonstrate that Neanderthals were more developed than originally thought.

Denisova hominins were first discovered in a remote cave in Siberia in 2010. Genetic decoding proved the fossil remains to be different to the Neantherthal and H.sapiens remains also found in the same cave. The bones were dated to 41,000 years old. The genetic evidence suggests Denovisans share a common ancestor with Neanderthals (likely H.heidelbergensis from Africa) and that they, too, interbred with modern humans and contributed their DNA to populations in Europe, Melanesia and Australia. In 2013, analysis of a 400,000-year-old bone found in Spain, which palaeontologists believed to be a specimen of H.heidelbergensis, revealed that it shared more DNA with Denovisans (who lived far away in Siberia) than with Neanderthals (who lived in the same region) or modern humans.

While the origin of these humans remains shrouded in mystery, it appears that each developed species of the Homo genus competed with the previous one, contributing to its disappearance. They may even be partially responsible for their own demise, as recently found Neanderthal remains in a cave on the Iberian Peninsula suggest that cannibalism was practised by Neanderthals during periods of food scarcity.

H. sapiens first appeared on the stage of life about 200,000 years ago, as the oldest fossil of this genus was found in Ethiopia, and its age estimated at around 195,000 years. They gradually spread around the globe, with their oldest-known artefacts dated to more than 120,000 years ago in the Arabian Peninsula, 70,000 years ago in India, around 55,000 years ago in Australia, 45,000 years ago in Europe, about 15,000 years ago in the Americas and only a few thousand years ago on remote oceanic islands, such as Madagascar, New Zealand and Iceland. However, these numbers are frequently changing in response to new scientific findings. While we, H.sapiens, are the only extant species of the Homo genus, our origins, time of appearance, ancestors and stages of development are matters of continuous debate—influenced by many different scientific, historical and ideological viewpoints. Nevertheless, much of the evidence used to prove opinions lacks certainty and could just as easily be explained by alternative hypotheses, leading to endless discussion and analysis.

The fact is that H.sapiens was equipped with the appropriate physical, mental, social, behavioural and psychological capabilities to gain dominion over uncountable other organisms that had gained a foothold on earth millions or billions of years before the first humans ever appeared. Since their appearance, they have undergone at least one genetic bottleneck resulting in a possible single origin—one woman—of all humans on earth today. Astonishingly, the genetic variability of mere 55 chimpanzees is twice the genetic variability of more than 6 billion humans alive today. Whether one or more, such bottlenecks took place about 100,000 years ago and led to a catastrophic reduction in the human population and probably the disappearance of pre-existing humans in most places as well. There were episodes of rapid climate change and probably active volcanism which occurred when humans were already scattered around the globe. Only those living in tropical climes and habitable environments were able to survive.

Fig. 6.4 Top: This satellite picture of the area to the south of Adam shows an old alluvial fan with winding wadi channels (in the centre), cut through and bounded by a younger fan with relatively straight river courses. The difference suggests that the older wadis flowed more continuously than those of today, reflecting a rainier climate in the past. Bottom: remains of flat, fine laminated sediments originally laid down in ancient lakes and later eroded by wind. (Pictures are from the book: The Geological Heritage of Oman by PDO)

6.3 Humans and Climate Change in the Last One Million Years

The Arabian Peninsula had been exposed to highly changing climates since one million years ago to date, ranging from pluvial periods, wherein rivers flew and grass grew, to cold and dry periods where sand had deposited and the fertile lands turned into arid deserts. Such contrast can be seen in the sedimentary strata of that period, as they range between rivers and lakes sedimentation and extensive sands. Such change may be directly linked to the earth's orbit around the sun, which, in turn, affects the sun rays that reaches the earth in what is known as 'Milankovitch Cycles'. A closer look at the map of Oman as it appears from the satellites today would reveal ancient lakes and voluminous rivers that were abundant in Oman just a few thousands of years ago (Fig. 6.4). This piece of land was once a favourable stop for the early humans to live in and to explore its sources, as it was not quite different from his homeland, Africa, and it

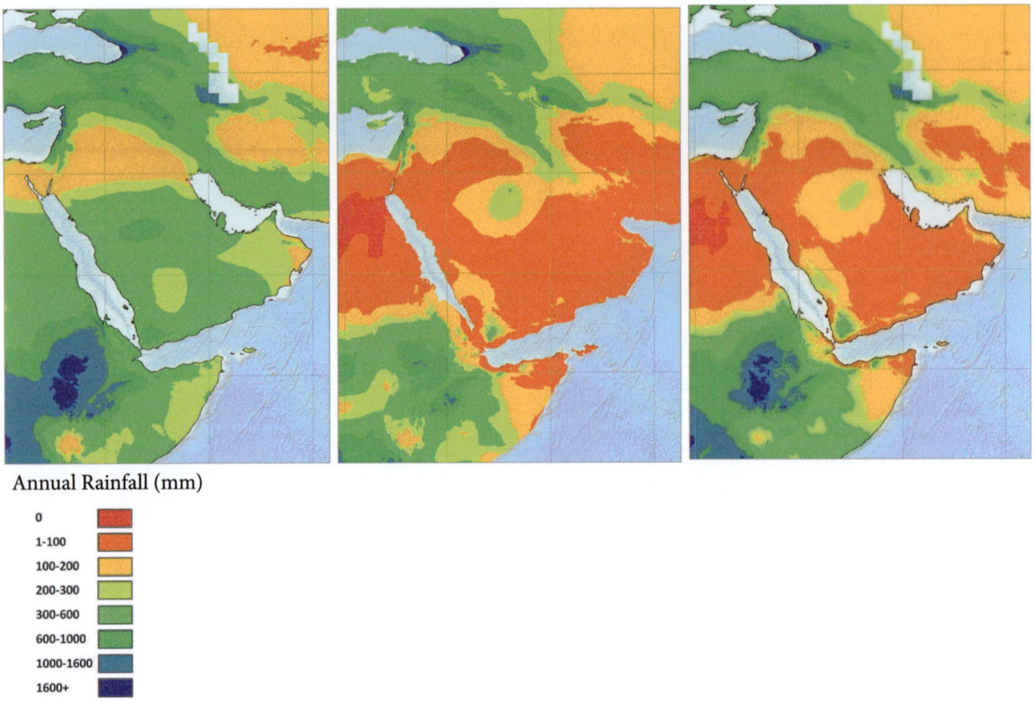

Fig. 6.5 Annual rainfall in Arabia in the last 125,000 years (source: worldclim.com). The Arabian Peninsula has turned into an arid desert in the last 100,000 years. The figure on the left is before 125,000 years, the figure in the middle is before 20,000 years, and the figure on the right represents the rainfall today. The more green or blue the area is, the more the annual rainfall it experiences. Today, most of Arabia receives less than 100 mm of rain per year, but it was even worse 20,000 years ago during the last ice age; 125,000 years ago, Arabia was savanna grassland and it would have been very attractive for humans to live in

was probably used by Homo erectus as a stop while crossing from Africa to East Asia. Archaeological surveys on the mud sedimentations around the marshes have revealed traces of the Homo erectus at the banks of these ancient lakes, as such lakes were the perfect place to lurk for the animals that would come near the water and to hunt and chop them by means of the stone tools that were abundant at the area. Additionally, the traces of Homo erectus were also found at relatively elevated areas, which were probably used by Homo erectus to detect the grass and water places as well as to watch out animals from atop. In short, man needed, along with all these natural factors, the raw materials in order to make his hunting and dismemberment tools.

During glacial periods, the earth's climate becomes cool and arid. Large areas of shallow seas become exposed to wind erosion because of a global drop in sea levels and vast areas of continents are covered by sand dunes. When temperature and precipitation increases during the interglacial periods (the warm intervals between the glacial periods), these arid deserts are often transformed into thickly vegetated areas that probably covered huge regions from east Africa to Europe and Asia. Therefore, the Arabian Peninsula represented a bridge for man's migration to Europe and Asia at the times when its deserts transform into pastures (Fig. 6.5), and perhaps different human species had moved to live there during the pluvial periods, while the same deserts play the role of an impervious barrier during the dry glacial periods, and, apparently, Homo erectus was moving generally around the Arabian Peninsula along the coastline, particularly during the dry periods, but when the rivers were filled with water, he would migrate

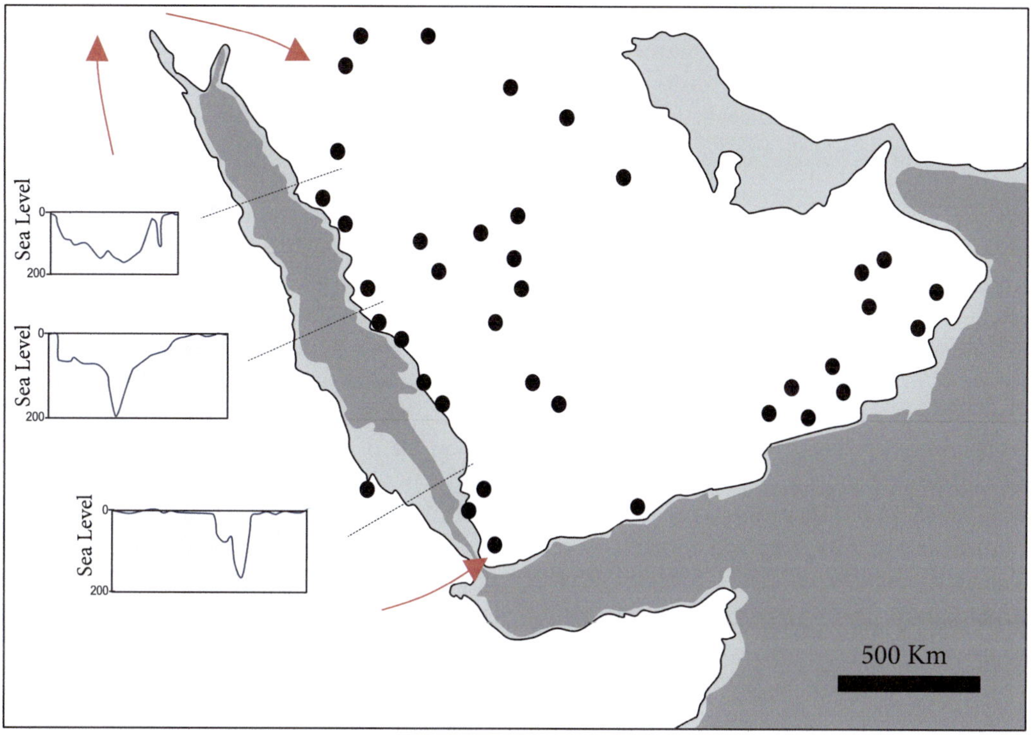

Fig. 6.6 Red arrows represent the possible routes used by early humans to cross from Africa to Arabia more than 100,000 years ago. The black dots represent the distribution of Lower and Middle Palaeolithic archaeological sites in the Arabian Peninsula (Redrawn from G. Bailey and C. Vita-Finzi, 2009). A number of ancient Palaeolithic sites exist in south and Central Oman. These are mainly localized around Wadi Darbat and the village of Muday in Thumrait, as well as around Al Huqf and Umm As Samim in Central Oman. The figure also includes on the left a number of transects across the Red Sea. Arabia and Africa remained detached along Bab Al Mandeb Strait even in periods of extremely low sea levels. The figure shows that greater part of the eastern coast of Oman was exposed above sea level (the pale grey colour); therefore, some archaeological sites might be today covered by shallow seas

through these rivers to the different vast areas inside and outside the Peninsula. The great variation in the Arabian Peninsula's climate had enriched the biodiversity as well as the lifestyle of the early humans. There is still enormous scope for research to understand the climatic changes during the glacial periods and how these changes affected the migration of humans within and across Arabia, and particularly, Oman.

Homo erectus had travelled from Africa to the Arabian Peninsula before about one million years ago and arrived at Oman around the same period, as some sources indicate that his traces may exist in Wadi Darbat in the Governorate of Dhofar. Other resources indicate the possible existence of Homo erectus tools in the Governorate of Al Wusta among other areas in the centre of Oman (Fig. 6.6). This migration had probably been made through Bab Al Mandeb straight, or perhaps the first human migration was northward through Sinai Peninsula, although it is possible that the migration happened through both of the routes, but the relocation through Bab Al Mandeb remains the more admissible hypothesis due to the presence of many archaeological sites of the ancient human at the south of the Arabian peninsula as compared to its north (Fig. 6.7). But for such migration to be possible, Homo erectus must had crossed the sea either by swimming, which is relatively unlikely, or by using primitive wooden rafts, given that this Strait can never be closed even when the sea had reached its lowest level at the Pleistocene glacial cycles, which is substantiated by the geological evidences found in the area. Nevertheless, the distance between the Asian and African continents through Bab Al Mandeb Strait would decrease to only few kilometres during the glacial periods, making the crossing of it easier for the Homo erectus. The distance today between the two shores is approximately 30 km (Fig. 6.7), and Homo erectus fossils were found at some of the Indonesian islands far from the land, which serves as evidence that Homo erectus did use primitive water rafts.

Now, the question arises as to why ancient humans decided to leave Africa in the first place. Why move to a distant area of a journey that involves many hardships and risks to reach? Perhaps the motivation then was the scarcity of resources and the fierce competition for food in Africa during the glacial periods of the Pleistocene epoch. Indeed, it is the cruelty and the tenderness of the earth that drives man to cling to or leave a certain land. Perhaps ancient man wanted to exploit the bounties of the sea due to the paucity of food on the mainland during the glacial periods. Or maybe it was simply the curiosity and the desire to conquer the farther shores that drove this creature to explore a new land far from home (Figs. 6.8 and 6.9).

It is worth mentioning here that the Afar area of Ethiopia, on the African side of the Bab Al Mandeb Strait, is one of the richest sites anywhere in the world for Hominini fossils. Nevertheless, none of these ancient species apparently crossed the sea until the advent of H.erectus.

Upon arrival on the Arabian Peninsula, H. erectus lived by hunting the different wild animals, such as elephants, wild horses and buffaloes, all of which were abundant during the pluvial periods. In times of hardship, however, it is likely they ate carrion—the carcasses left by large predators or which died through age or disease. Most likely, H.erectus habitually ate such meat raw, and the lack of cooking could well have lead to deaths among tribes—either from bacteria in decaying meat or through lack of nutrition from badly digested food. H.erectus also had to deal with the existence of predators that would either compete with him in hunting other animals or try to devour him as prey. As a

Fig. 6.7 View across the Bab Al Mandeb Strait, looking east (photo by Tony Farmington). There is a deep seaway along the strait, so at no point in history has dry land ever connected Africa and Arabia. Nowadays, it is about 30 km across

Fig. 6.8 Various ecosystems in Oman have supported the settlements of humans in different climatic conditions, from top left to bottom: Wadi Al Arbiyeen, Bar Al Hikman, Muhut lagoons, Ash Sharqiyah Sand, Musandam and Dhofar

result, H.erectus tried to take advantage of the complex topography of the area and enhance his competitive edge in terms of planning to catch prey and hiding from predators. Such useful landforms were available thanks to the mountain chains in the south and west of the Arabian Peninsula. Besides being perfect hiding places, the mountainous terrain held natural reservoirs of water, which would attract wild animals to live nearby. H.erectus would also have taken

Fig. 6.9 Acheulian hand axe, up to 15 cm long. Photo by Didier Descouens

advantage of the bounties of the sea, consuming fish and seashells for his subsistence, and probably settled in the coastal areas of the south and west of the Arabian Peninsula, although archaeological evidence for this is both insufficient and inconclusive as yet.

During his stay in the south of the Arabian Peninsula, H.erectus had, during the period between 500,000 and about 120,000 years ago, been affected by the monsoons of the Indian Ocean, which brought many benefits during the pluvial periods and attracted a mix of early humans who migrated and lived there in such a way to create a genetic mix with substantial biodiversity. Cave sedimentations in both the south and north of Oman indicate that such pluvial periods reached their peak between 330,000 and 300,000 years ago (Fig. 6.10).

The large number of stone tools from that period clearly indicates the proliferation of early humans in southern Oman, and as time went on, such tools became more diversified and were used for multiple purposes, becoming lighter and sharper. Interestingly, fossils of H.erectus have not yet been found on the Arabian Peninsula, although the artefacts he left behind—the stone implements and flint tools—abound on the outskirts of wilayats Thamreet and Salalah, at what is known as the plateau of Najd, and more scarcely in the Al Huqf area on the eastern coast of Oman. Such traces are lacking in the northern mountains in Oman, which may be due to the deserts that separate the north from the south blocking the migratory routes north.

The period in which H.erectus lived is known as the Acheulean Period, and the stone tools he used take the same name. They derive from Saint-Acheul, an area near Amiens in northern France where the distinctively shaped tools were first identified in the mid-nineteenth century. They are relatively large stone tools, ovoid in shape like a pear or an egg, and were developed

Fig. 6.10 Climatic change in the Pleistocene (last 650,000 years), blue is cold and red is warm. (after Tomruen, 2011)

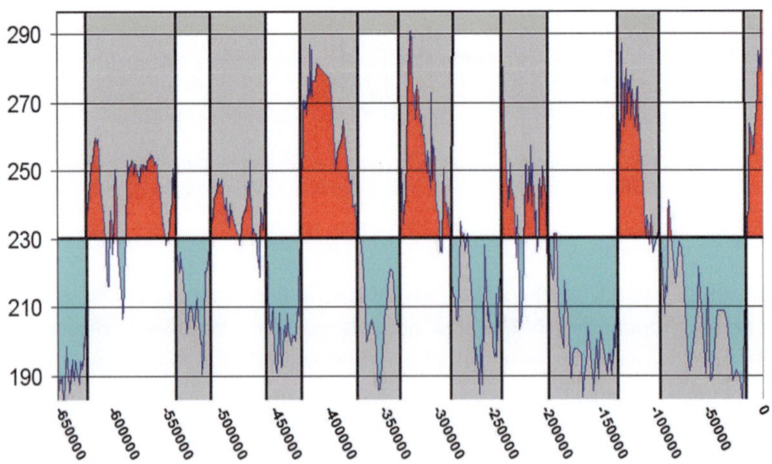

by early humans between 1.7 million and 100,000 years ago. They were used as hand-held axes for hunting animals, trimming wood, digging roots and so on, and differ from later stone tools in that all the faces of a stone were flaked or sharpened, rather than chipping small pieces of a larger stone and making tools from the flakes (Fig. 6.9).

Dating ancient stone tools like these is extremely difficult, as they are highly durable, and ancient styles often persisted into relatively modern times across different regions. For archaeologists specializing in the Arabian Peninsula, the study of such implements warrants further investigation and wider geographical linking.

There is no fossil or archaeological evidence of Neanderthals in the south of the Arabian Peninsula, although the theory that they moved southwards from their northern home ranges into the Levant and Iraq is supported by fossils north of the Arabian Peninsula and in Iran. It may well be that the soaring temperatures of the southern reaches of the Peninsula kept the heavily built Neanderthals—adapted for colder climates as they were—from straying too far south. Conflict or competition between H.erectus and later with H.sapiens may also have resulted in different colonies keeping apart from each other.

6.4 Climatic Changes and Homo Sapiens Distribution in Oman in the Last 200,000 Years

Homo sapiens first appeared about 200,000 years ago, and since then has continually explored new places to live. Mankind was endowed with certain physical and mental characteristics that enabled him to travel and settle in a better way than his ancestors, and for him, the southern Arabian Peninsula, with all its bounties in the coastal and mountain areas, represented an attractive proposition for settlement (Fig. 6.11). It was probably for this reason our ancestors crossed the Bab Al Mandeb Strait from Africa to Arabia, and H.sapiens apparently arrived in Oman after the last but one glacial period which ended about 150,000 years ago (Fig. 6.12). The oldest evidence for settlement in Oman—stone tools—dates back to around 106,000 years ago and was found in the mountains of the Dhofar Governorate. Optically stimulated luminescence (OSL) dating was used to determine the age of these tools. OSL technology can calculate the length of time since an artefact containing minerals such as quartz was last exposed to sunlight, making it useful for dating stone artefacts. Similar tools were found in the Nuba Mountains, north of Sudan, in the 1960s. They date back to

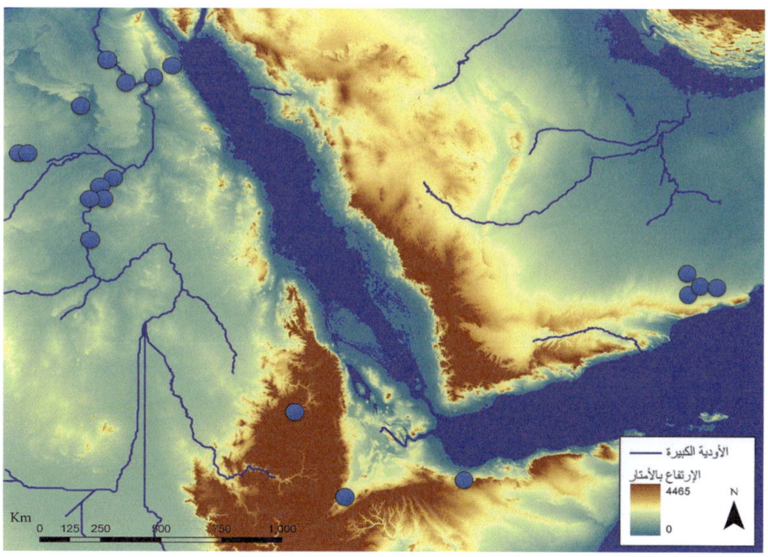

Fig. 6.11 Sites of Nubians in Arabia and Africa discovered so far (after Jeffery Rose). The sites in Dhofar are dated to 106,000 years ago, whereas those in Africa date as far back as 125,000 years and more

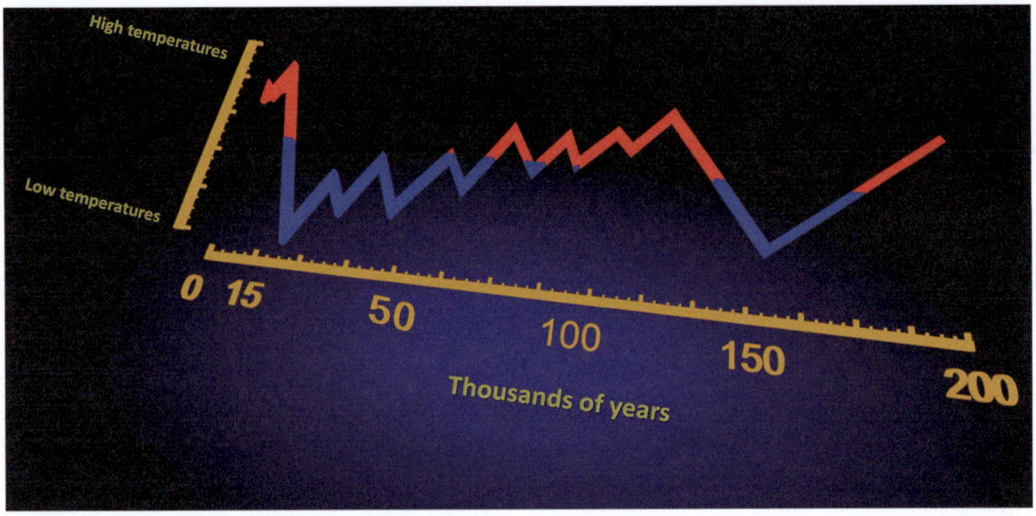

Fig. 6.12 Glacial periods in the last 200,000 years. Two major glacial cycles occurred around 150,000 and 20,000 years ago. Humans probably migrated to Arabia about 120,000 years ago after an extreme glacial event and during the start of a pluvial period that experienced high rainfall. In the last 200,000 years, Arabia most likely experienced three extremely arid periods with low rainfall, around 70,000 years ago, 20,000 years ago and today

around 125,000 years ago and are termed 'Nubian Tools' (made by Nubian Man). They are distinguished by being sharp, three-faceted tools, with one blunt edge and one un-knapped oval edge which fit into the hand. Evidence such as this suggests that Nubian Man crossed from Africa to the Arabian Peninsula following in the footsteps of H.erectus who went before (Fig. 6.13).

Interestingly, one possibility could be that our modern H.sapiens first appeared in the southern Arabian Peninsula and then migrated to Africa. This means the first migration would have been from inside the Peninsula and not the other way around. There is evidence of successive generations of human presence in some locations in Dhofar for a period exceeding 500,000 years using similar techniques to make tools (Fig. 6.13). However, the similarity of tool-making methods alone is not sufficient to determine the age of the tool, especially when the distance between two archaeological sites is extended.

Many anthropologists believe that there were multiple migrations between Africa and the southern Arabian Peninsula through the Bab Al Mandeb Strait due to repeated climate changes on either side of the Red Sea. Some parts of the Peninsula were almost devoid of human presence—particularly between the Pleistocene and Holocene epochs, or even earlier—as indicated by genetic studies. This phenomenon is apparently linked to the last glacial period, and the Arabian Peninsula was perhaps the destination of wholesale migration at the beginning of the Holocene epoch. Many areas have so far remained unexplored, particularly in and around the Empty Quarter, which, while buried under sand now, may have once held the camps and tool-making sites of early ancestors.

The earth had witnessed two glacial periods since the emergence of H.sapiens (Fig. 6.12). The first ended around 150,000 years ago, while the second ended about 15,000 years ago. The temperatures dropped drastically 20,000 years ago, while a pluvial climate prevailed before and after. The ebb of the sea left the Arabian Gulf as a low-lying flatland divided by the two rivers of Euphrates and Tigris whose estuaries were near Musandam. During the pluvial periods, the rivers and lakes were abundant in the Empty Quarter desert as well as in the middle of Oman, reaching

Fig. 6.13 Flint artefacts in Dhofar (photo by Dr. Jeffery Rose)

a peak sometime before 70,000 years ago. During the climate change and temperature fluctuation in the Arabian Peninsula, the sea level ebbed and flowed, and wind speeds increased and decreased. During the dry glacial periods, the sea level fell to around 120 m below current levels. In contrast, during the warm pluvial periods, sea levels rose by tens of metres above current levels.

Generally, the banks of rivers and river valleys are the optimum place for living, particularly during the pluvial periods, due to the abundance of pastures and game, even in areas which look like deserts today. Plankton, on which fish feed, increased with the monsoons, bringing bountiful harvests in rivers lakes and the coastal water. Such climate conditions encouraged multiple migrations, especially with the impressively diverse terrain of the Arabian Peninsula in general, and in Oman in particular (refer to Fig. 6.8), which made it available for man to migrate according to the prevailing climate of the season or period. There were certain areas in Oman that provided man with a safe haven to live in for more than 100,000 years without the need to migrate for long distances. This includes the Dhofar Mountain Chain and the Al Huqf Mountains in eastern Oman. Both provide year-round natural shelter and abundant resources, allowing ancient peoples to escape the hardships of inter-glacials.

The mountains of Dhofar can be divided into two main parts. The first is an elevated and wide eastern mountain chain that drops abruptly into the Indian Ocean and provides a block against the summer monsoon that sweeps across the ocean from the south-eastern part of Asia. The mountains are covered in lush, green vegetation which thrives in the mists of khareef and results in a uniquely cool and wet summer—very different from all other parts of Arabia (Fig. 6.14, top). The second (Fig. 6.14, bottom) is represented by extensive western hills that gently dip towards the Empty Quarter. These hills are incised by ancient valleys that run from east to west and in the past, probably

Fig. 6.14 Dhofar Mountains. Above is the eastern side and below is the western side (bottom photo is by Jeffery Rose). This region of Oman was most likely highly populated and today's dry streams were full of running water

flowed with abundant rivers, especially during the rainy season. Yes, this is only half the story, because the western hills of Dhofar are also very rich in flints or chert, which were essential raw materials for ancient peoples. They used such rocks to make tools —arrowheads and spears for hunting, axes and awls for working wood and bone and scrapers for cleaning hides and cutting meat.

Anthropologists and geologists more or less agree that the Dhofar Mountains have hosted sequential settlements of humans since the Middle Palaeolithic. These settlements are reflected in the different shapes of stone tools found in many places in Dhofar. There are at least 250 archaeological sites used by H.sapiens or modern humans in Dhofar. The most remarkable are dated to more than 100,000 years ago and belong to a type of known as the Nubians. Nubian tools are widespread around the village of Muday in the Wilayat of Thumrait and extend to at least the beginning of the Empty Quarter on the western limit of this Wilayat (Fig. 6.15). The extensive distribution of these artefacts indicates long and prominent settlements of Nubians. Muday with its wide plains, water-rich valleys and flowing springs seems to have been a haven for humans, particularly during glacial periods, when annual rainfall declines and the environment becomes harsher in the high mountains.

The Al Huqf area in the east of Al Wusta Governorate is another strategic place for ancient man to live in. It contains small hills with a rich geological record and different types of rocks that enabled man to exploit the environment in different ways (Fig. 6.16). The area is open to the Indian Ocean, and its coasts are among the richest in terms of fisheries in Oman. It also had fertile valleys with copious waters flowing during the pluvial periods. Examination of mud sedimentations indicate that there were ancient lakes in this area and they stand today as witness of the presence of more than 350 archaeological sites and millions of different stone tools in an area which extends over 200 km from Wilayat Mahut in the north to Wilayat Duqm in the south. The enormous number of archaeological sites and different stone tools indicates human occupation for long successive periods and might be an evidence of the high population of this area during the Palaeolithic era. Some researchers believe that there are even more ancient archaeological sites in the Al Huqf area which are now submerged following the increase in sea levels. It is possible that the long and continuous use of the area resulted in a depletion of the vegetative cover and other natural resources, leading desertification and the subsequent migration of most of its population towards the north and south of Oman.

The stone tools discovered in the Al Huqf area are very diverse in terms of shape and age, varying from one location to another, and although accurate dating is a controversial issue, they most likely belong to the Palaeolithic era and include arrowheads, scrapers and pestles (Fig. 6.17).

A warm and pluvial climate prevailed in the world between 9,000 and 5,000 years ago at the beginning of Holocene epoch, and this phenomenon is known as 'Holocene Climate Optimum'. This period was characterized by high sea levels as the sea covered large parts of the land. For example, the water of the Arabian Gulf reached the ancient city of Ore south of Basra city. At the same time, the glaciation level at the two poles decreased due to higher temperatures by several degrees. This phenomenon had apparently started hundreds of years before that point, as cave sediments in Dhofar Governorate indicate that the stalactites and stalagmites had formed prior to 10,000 years ago, while the rocks were deposited in the caves of northern Oman around 9,000 years ago.

6.5 Climatic Changes and Human Civilizations in Oman Since 9,000 BC

Mankind settled in Oman throughout the ages. The Hafit Period (3,200–2,700 BCE) and Umm an-Nar (2,700–2,000 BCE) were identified during the early Bronze Age and Wadi Souq (2,000–1,300 BCE) during the late Bronze Age and Lizq (1,100–600 BCE) and Samad during the Iron Age (Fig. 6.19). These periods are generally

Fig. 6.15 Big assemblage of Nubian artefacts (5–10 cm long), collected by Dr. Jeffery Rose and his archaeological team in the vicinity of Muday in the Wilayat of Thumrait. These artefacts were probably left behind by the first humans that migrated to Arabia. The figure below represents a pyramid of artefacts (each about 10 cm long), with the oldest at the base (more than 200,000 years old) and the youngest on top

named after the place where evidence of distinct phases of human occupation were first discovered. In the following texts, we will cover only the civilization periods that thrived in Oman since the 9th millennium BCE until the end of the Iron Age or prior to the advent of Islam in Oman. This land has been inhabited by humans since prehistoric times, as evidenced by the huge number of archaeological sites here. They reveal a rich history of the existence of skilled hunters,

Fig. 6.16 A typical landscape of Al Huqf area, as represented by Wadi Gharif in the Oryx Sanctuary in the Governorate of Al Wusta. Many artefacts can be found on the hills of these wadis. Ancient people utilized the flints of the Permian rocks to make their tools

Fig. 6.17 A selection of heavy artefacts from different sites in the Al Huqf area in Central Oman. The left sample is about 15 cm long

hard-working metal prospectors and experienced manufacturers. This tradition still exists today. Over time, Omanis have established strong cultural and commercial relationships with the whole world.

About 10,000 years ago, the earth had entered a warm phase. Cave sediments from Dhofar Governorate indicate that the stalactites and stalagmites formed over 10,000 years ago, while in the caves of northern Oman, rocks were deposited 9,000 years ago. This could be attributed to a gradual northward shift of the centre of the monsoon coming from the Indian Ocean from latitude 15° to latitude 25°. Temperatures at the poles rose, glaciation decreased, and the 'Intertropical Convergence Zone' expanded. A warm and pluvial climate prevailed between 9,000 and 5,000 years ago at the beginning of the Holocene epoch—known as 'Holocene Climate Optimum'—characterized by warmer temperatures and higher sea levels. The waters of the Arabian Gulf reached the ancient city of Ore to the south of Basra city, now some 100 km inland.

During this time, people spread all over the Arabian Peninsula as vast areas turned into grasslands full of Acacia, Prosopis and Tamarix trees. Lakes formed and the hospitable environments of the deserts, mountains and plains enabled man to form settled communities. The emergence of civilizations and human advancement was facilitated by the climatic conditions and the unique geographical position of the region, central among the three major continents: Africa, Asia and Europe. Migrations between north and south became commonplace.

Such favourable conditions, however, did not last for long. Around 6,000 years ago, temperatures began to fall, rainfall decreased, and droughts became more frequent. The monsoon receded towards the south, and around 5,000 years ago, the lakes that once filled Oman and Arabia evaporated. The vast riches of Oman's natural resources were also being depleted. The smelting of ores, such as copper, resulted in the loss of thousands of trees, eventually contributing to the desertification and deforestation of the land. Since those days, temperatures have not markedly changed in Oman, nor have sea levels altered, although the amount and frequency of rainfall have continued to diminish and drought has gradually increased.

6.5.1 Neolithic Era (New Stone Age)

At the beginning of the Holocene epoch after the 9th millennium BCE, the population of Arabia dramatically increased. Since the numbers of wild animals had already been depleted due to overhunting, communities began the transition from hunter-gatherer to farmer, with small settlements practicing primitive agriculture and domesticating animals. This shift brought about unprecedented changes. It was an entirely new lifestyle, less dependent on the environment, the vagaries of weather and the migration of livestock. Societies became more coherent, developing their own traditions and religious identities, wherein each individual, man or woman, had a particular role and its attendant rights and responsibilities.

The oldest archaeological sites in Oman from the Holocene go back as far as 9,000 BCE in the Wattiyah area of Muscat Governorate. Apart from this, no sites older than the 7th millennium BCE have been discovered. As time went by, the primitive communities spread from north to south, with the emergence of the societies that relied on mass hunting of large animals, such as zebra, ibex, gazelle, wild camels and ostrich, as depicted in the ancient drawings in Wilayat Shinas in the North Batinah Governorate and Wadi Watiyyah in Muscat Governorate. Around the end of the pluvial period, the number of large animals had decreased due to the climate changes as well as excessive hunting, and they no longer played a role as an essential food source. Instead, humans became more dependent on sedentary agriculture, animal domestication and fishing, although some communities had a semi-nomadic lifestyle, moving with their domesticated animals between the coast in winter and the mountains in summer. At about 6,500 BCE, such settlements began to use pottery which was partially brought from Mesopotamia. Such pottery is known as 'Slave Pottery', and some scholars call this time the period of slaves.

The earliest coastal archaeological evidence dates back to the end of the 6th millennium BCE and is located at sites such as Ras Al Had in Wilayat Sur, as well as at Al Sweihat in Wilayat Jalan Bani Bu Ali and Ras Al Hamra in Wilayat Baushar. These sites were inhabited by humans during long and successive periods throughout history due to their abundant natural resources.

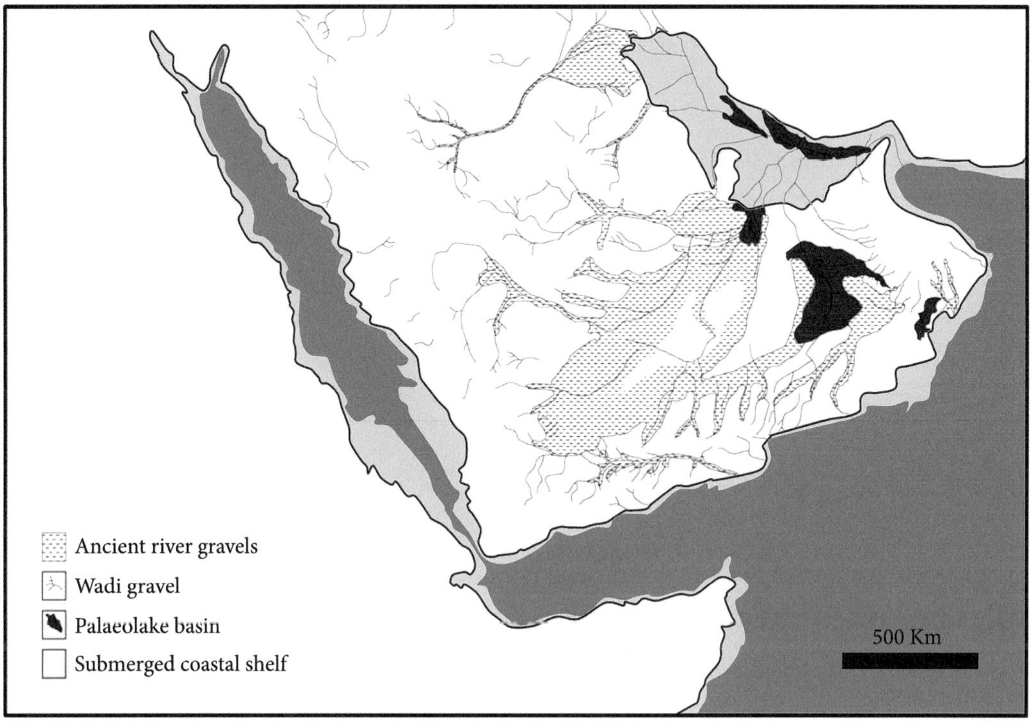

Fig. 6.18 Map of ancient drainage channels, alluvial deposits and paleolake basins throughout the Arabian peninsula during pluvial times. The boundaries of the continental shelf indicate the extent of Arabia during periods of reduced sea levels (roughly between 75 and 8 thousand years ago) (Rose and Usik, 2009). The global sea level was reduced by around 120 m. During these periods, the floor of the Arabian Gulf became exposed and many rivers (including the Tigris and Euphrates) discharge water close to Musandam in the Sea of Oman. The Arabian Gulf must have been a land with many streams of water, and therefore, it attracted many people to live within it. Throughout most of the Upper Pleistocene and early Holocene, a considerable amount of runoff in southwest Asia was funnelled into the Gulf basin via submarine aquifers flowing beneath Arabia. The Tigris and Euphrates Rivers were also flowing from the Anatolian Plateau. Moreover, the drainage network discharged water from the Zagros Mountains in Iran to the south. All of these systems converged in the centre of the Gulf basin, forming what is known as the Ur-Schatt River, which ran through a deeply incised canyon that is still evident in the extant bathymetry (Seibold and Vollbrecht, 1969; Sarnthein, 1972)

Such communities exploited the shallow seas and coastal marshes for fishing and collecting shells. On the coasts of the Indian Ocean, they were frequently unable to put to sea in the stormy conditions, or venture far from shore even in calmer weather, so fishing was a limited occupation. For communities along the Arabian Gulf, however, life was more predictable, as the waters were calmer and not so deep or dangerous. Reed boats waterproofed with some form of tar were in use by coastal communities from the 5th millennium BCE.

During the 4th millennium BCE, political and social transformation took place in Arabia, with the dawning of the most important civilization centres worldwide. Writing was invented around 5,000 BCE, trade and supplies increased, economies developed, the pottery and textile industries thrived, and land was turned over to agricultural use. The hierarchy of society changed and became more vibrant. Microstates began to emerge, and commercial and diplomatic exchange became more common. Oman, however, stood on the brink of a similar transition, but was not yet fully engaged. Archaeological evidence points to the emergence of coastal communities that used the raw materials available in their surroundings, such as shells and

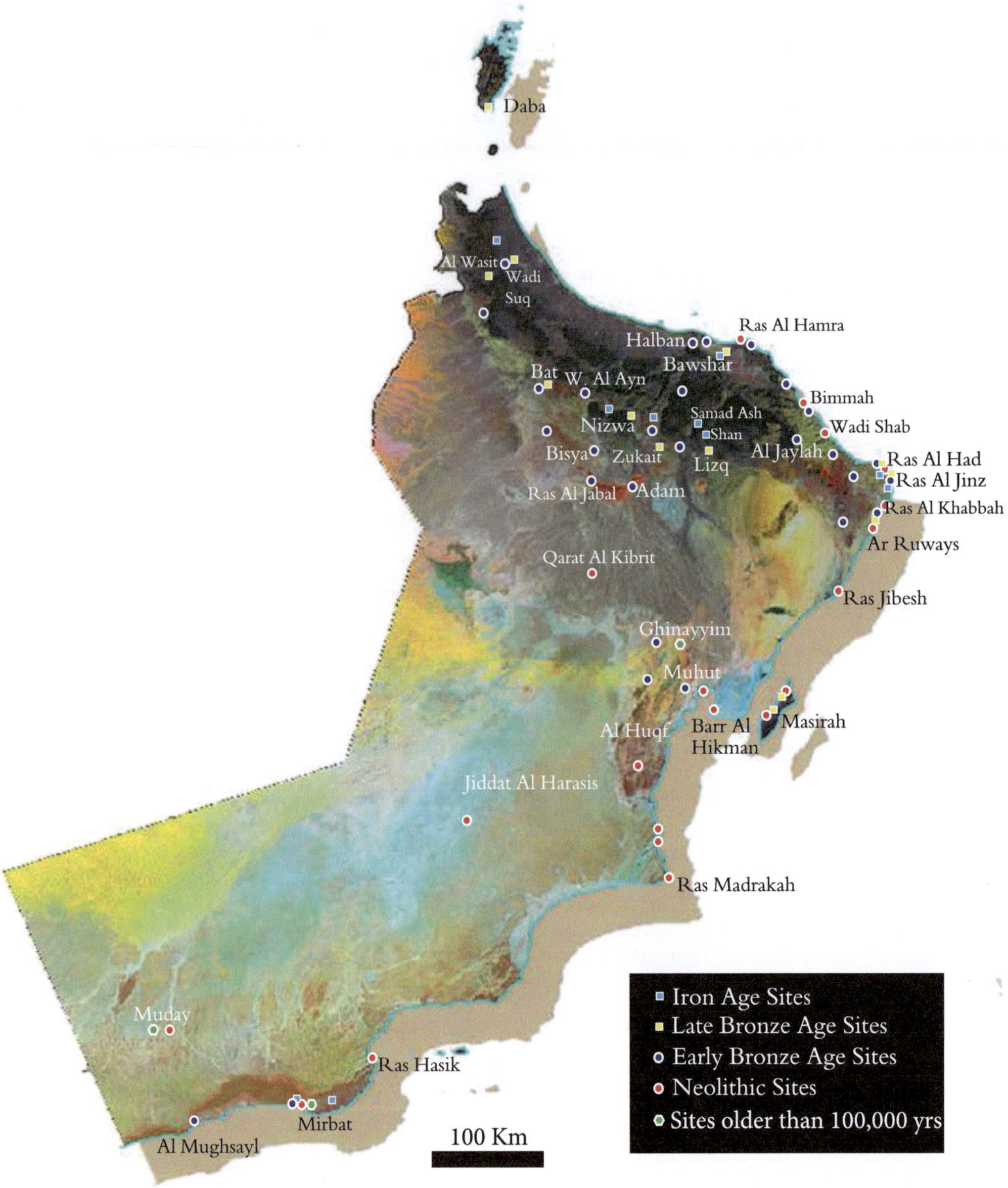

Fig. 6.19 A map of Oman showing the locations of important archaeological sites from various periods. Some of these sites (e.g. Wadi Darbat in Mirbat) contain acheulean tools produced during the Lower Palaeolithic, more than 200,000 years ago, and were made by Homo erectus, before the arrival of Homo sapiens. The oldest-known artefacts of Homo sapiens in Oman are from the area around Muday near Thumrait in Dhofar. They are estimated to be about 106,000 years old. Flints are scattered nearly everywhere in this area. Similar tools have also been found in the northern part of Al Huqf in Central Oman. Neolithic sites are mainly concentrated in the eastern part of Oman, particularly around the headlands such as Ras Al Had, Ras Madrakah, Ras Al Hamra and Ras Hasik (Ras is Arabic for headland) marking the exploitation of the sea and coastal lagoons. However, the distribution of archaeological sites might also be biased by the concentration of archaeological work. The locations of the early Bronze tombs (from the Hafit and Umm an-Nar Periods) are largely scattered across northern Oman, indicating dense populations of farmers, herders and fishermen at the dawn of the Arabian civilizations. These tombs are collective family graves. During the late Bronze Age (the Suq Period), the number of settlements and burial sites decreased, probably indicating destabilization and the collapse of trade networks. As a result, this period is often known as a dark age. People settled in protected areas like Masirah Island. Later, in the Iron Age, a revival of development occurred leading to significant changes in culture and materials used

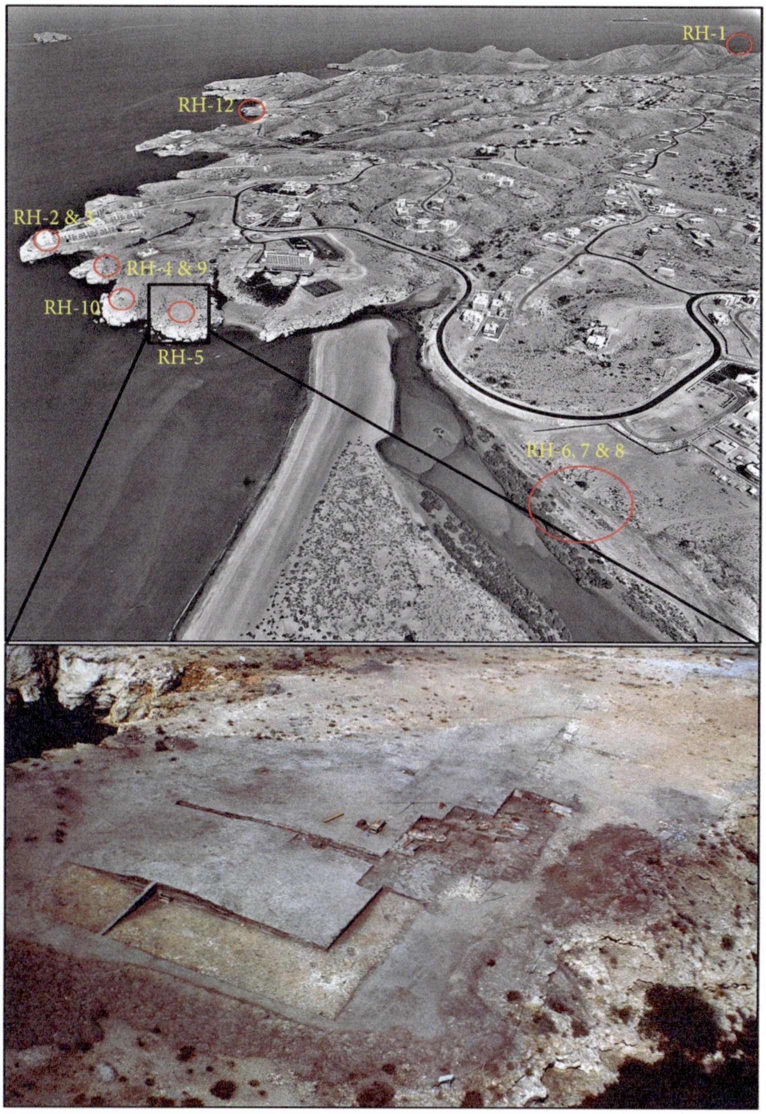

Fig. 6.20 Ras al Hamra. Top, an aerial photograph of the Ras Al Hamra area, taken in 1977.The red circles show the locations of the main archaeological sitesin this area, along the coast and where Wadi Aday meets the sea in alarge mangrove swamp of Al Qurum. Sites RH-5 and RH-6 are among theoldest archeologist sites in Oman and they are now protected. Thelower aerial photograph was taken in 1984 and it shows the earlyexcavations in RH-5. The site revealed one of the oldest villages inOman with its cemetery and yielded a lot of information about theculture and materials of the ancient fishing communities in Oman.(Photos: Ministry of Heritage and Culture.)

stones, in making their tools, but there is no evidence that such communities were trading with others at the time. Evidence from the 4th millennium BCE has been found at Ras Al Hamra in Wilayat Baushar and in the area located between the north of Sur and Al Ashkhara in Ash Sharqiyah South Governorate.

Ras Al Hamra-5 site is one of the premier archaeological sites in Oman as it contains hundreds of tombs, dwellings and food stores which contain large quantities of fishing tools, toiletries and the remains of foodstuffs (Fig. 6.20). The people of the past had carefully selected the location, as it provided everything required to support life in terms of land and sea. It sits at the point where Wadi Adai converges with the sea along Khor Al Qurm. Man had perhaps travelled along the valley to live among the mountains in search of water and pastures during the scorching summer periods. The oldest human fossils in Oman are here, dating back to the 4th millennium BCE around 5,500 years ago. Its tombs contain shells used as decoration and the bones of sea creatures, such as turtles, as though the people were preparing for the Afterlife by burying some of their belongings with them (Fig. 6.21). Such people were influenced by the marine creatures living around them. Some tombs even contain a group of 34 pebbles, which is the same number of eggs laid by a turtle at one time, and they often reconstructed the remains of their deceased in the shape of a turtle, as though they thought people would be reincarnated in turtle form. Groups were probably genetically isolated, possibly leading to a decrease in life expectancy to under 30 years. The remains of sharks and fish hooks found in the area indicate that the ancient residents of Ras Al Hamra plied the waves in wooden fishing boats and used fishing rods to catch fish from atop the hills where they used to live.

Among the important archaeological sites that date back to the end of the 5th millennium BCE and the beginning of the 4th millennium BCE is Ras Al Khabbah-1, located in Wilayat Jalan Bani Bu Ali. It is similar in many ways to Ras Al Hamra-6, which dates to the same period. Situated 30 m above sea level, it is bounded on the west by a saltwater creek (Fig. 6.22) and is in a prime position. At that time, it appears people preferred to live on headlands projecting into the sea due to their strategic location in attracting plankton and marine life such as turtles and fish. It was a unique fishing community that used to settle on the coast, perhaps moving up into the mountains during the summer months when the fierce tides reduced the fishing catch. A few bones of domesticated animals were found in Ras Al Khabbah-1 where people probably used to live in simple round houses as evidenced by the multiple round trenches found in the site. Also among the finds were a large quantity of stone tools including grinders, pounders and fishing net weights. An interesting find was a set of flat pieces of stone connected by a thread probably used as a primitive musical instrument during celebrations (Fig. 6.23). The stone artefacts also included 12 stone splinters that formed a sandstone jar considered as one of the oldest domestic vessels in Oman. Other finds include sharks' teeth, pierced twice in the base so they could be attached to poles and used as spears and fishing equipment including bone fish hooks and stone weights for holding down nets and lines. Beads and jewellery made from shaped bones and shell were also found. In general, Ras Al Khabbah-1 site constitutes an old community that lived on coastal fishing and collecting food from nearby tidal salt marshes (Fig. 6.24).

By the end of the 4th millennium BCE (about 3,100 BCE) and the beginning of the 3rd millennium BCE, the communities that lived at Ras Al Had, Ras Al Jinz and other coastal regions of the Governorate of Ash Sharqiyah South had evolved and expanded significantly in comparison with the communities at Ras al Hamra. In fact, the natural conditions prevailing for thousands of years in these coastal areas and the surrounding mountains created a sort of harmony between the native plants and animals live on the one hand and men on the other (Fig. 6.25). More importantly, the ecological structure of the area, even with the end of the monsoon season, was able—until the Holocene—to facilitate important urban development. It is worth mentioning here that Ras Al Jinz and Ras Al Had are Oman's richest fishing grounds throughout the entire year. This is due to the summer monsoon winds that blow parallel to the eastern coast of Oman from the south to the north (Fig. 6.26). They push the warm sea water towards the ocean to the east generating water currents between the deep, cold bottom of the Indian Ocean and the warm surface waters in a coastal upwelling. With the upward movement, huge amounts of organic material and mineral deposits from the bottom of the sea are dragged to top. These help the phytoplankton prosper near the surface of the sea which leads to the propagation and diversification of fish along the eastern coast of Oman. However, the big waves and rough water that often accompanies the high winds in the summer makes sailing dangerous. In the winter, large schools of fish migrate from the south to the north. It is Oman's fortunate location (truly Arabia Felix) that has endowed its mountains and coasts with such geological, archaeological and ecological richness.

The big shift witnessed by these fishing communities at the end of the 4th millennium BCE and the beginning of the 3rd millennium BCE was the vital step in becoming economically,

Fig. 6.21 Bones and tools recovered from Ras Al Hamra RH-5 dating to the 4th millennium (Joint Al Had Project, Ministry of Heritage and Culture). Top: A grave containing the skull of a human and the skull of a turtle. Most graves consist of molluscs, fish bones and turtle bones and eggs. In some cases, the human remains are arranged to imitate turtle skeletons, probably because people then believed that the deceased would be transformed into turtles or other totem animals. Centre: The skeleton of a young woman with a bracelet and head dress. Bottom: a selection of tools and jewellery recovered from graves at the site (RH-5) including (from left to right) a ring, a leaf-shaped earring, a shell pendant from a necklace and a stone tool. Beneath them are two views of an almost complete pot dated to 3,400 BCE. The pot was most likely imported from Iran and used to heat bitumen brought from Mesopotamia to waterproof fishing boats. (Photos: Italian Archaeological Mission to Oman.)

Fig. 6.22 Ras al Khabbah. Top: a number of circular trenches from Ras al Khabbah-1 probablydating back to around 4,000 BCE. They are often associated withpostholes and storage pits. They can also be found at Ras al Hamra(RH-5). The photo is from the Joint Al Had Project, Ministry ofHeritage and Culture. Bottom: a number of traditional huts in theNiger. Similar huts were used until recently in Oman, particularlyin the highlands of Dhofar. (Photos: Joe Ruhinski.)

technologically and culturally linked to the rest of the communities of the Arabian Peninsula. The eastern coast settlements of Oman as shown in Ras Al Had and Ras Al Jinz became more organized in terms of urban planning. Areas of domestic dwellings and burial ground began to be separated with distinct mortuary areas where the dead were buried under domes made of stone. Hundreds of stone cemeteries have been found around the settlements. Many of the previous stone and shell tools were still being used, but a new element had been added: metal. Archaeological sites belonging to this period are full of hundreds of copper fishing hooks in addition to fishing weights and thousands of fish bones. There are also thousands of shards of ancient pottery and pieces of decorative beads that used to be made of shell or stones (Fig. 6.27). This transitional time is sometimes referred to as the Chalcolithic Age—the bridge between the Neolithic and the Bronze Age. It is the period during which men used copper without mixing it with tin—prior to the

Fig. 6.23 Graves at Ras al Khabbah. Left: two burials from KHB-1 site dating to the late 5th millennium BCE. Next to each is a reconstruction of the grave with the layout of the body during burial. The top part of the body is missing from the lower grave. The top grave is for a young woman and the lower is for a mature man. The artefacts in the middle were recovered from the grave site: a bead, a tool some 7 cm long, which would have been threaded on a string and used as a musical instrument, a fragment of stone vessel and a shark's tooth with two perforations to be used as an arrowhead. (Images: Joint Al Had Project.)

discovery of bronze, which is a stronger alloy of copper and tin. Over time, the use of copper became more popular with the appearance of copper knives, chisels and pegs. There are also indications that an increased population demanded bigger fish catches and that as nets grew larger, so did the stone net weights these people left behind.

6.5.2 Bronze Age

During the early Bronze Age at the beginning of the 3rd millennium BCE (about 2,900 BCE), a group of settlements appeared along the Hajar Mountains from Musandam to Ras Al Had (see Fig. 6.19). These regional settlements came into

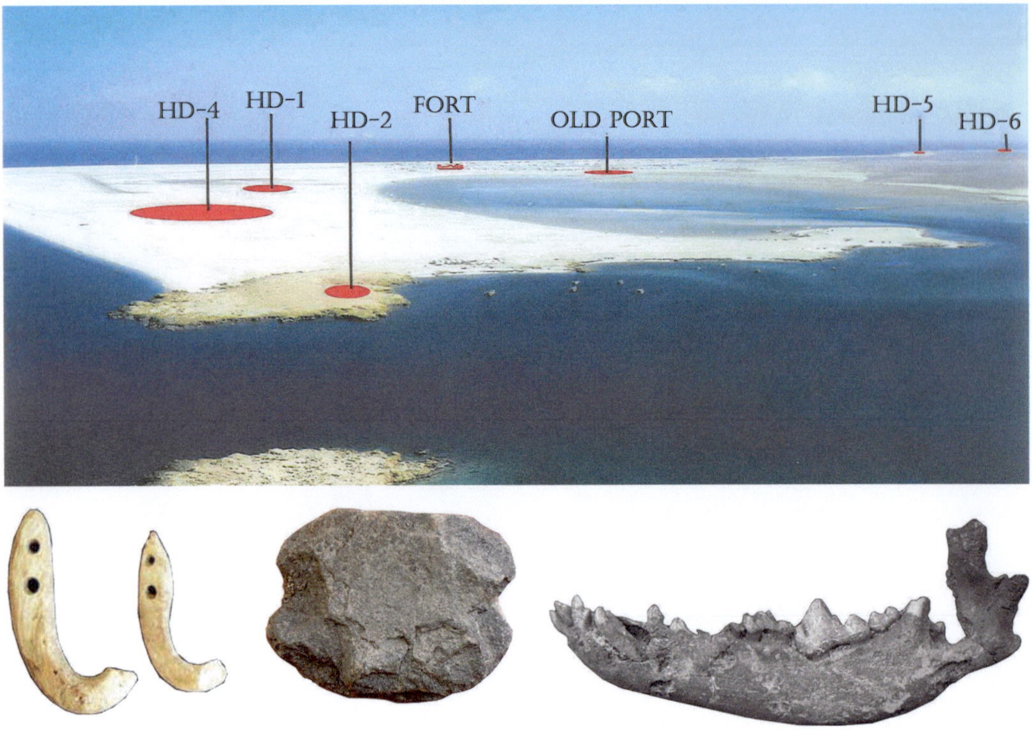

Fig. 6.24 Top: archaeological sites along the sand spit and lagoon of Ras Al Had. Sites HD-5 and HD-2 are from the Middle Holocene (Neolithic), HD-6 and HD-1 are from the early Bronze Age and HD-4 and the fort are from the Islamic Period. The figures below include from left to right two shell hooks, a net sinker and a lower dog jaw with cut marks—possible evidence for butchering and meat processing by humans. (Images: Joint Al Had Project.)

prominent contact with other Arab communities and were a sign of the emergence of large tribal organizations in established agricultural areas. They were characterized by the appearance of 'beehive' tombs between 3,200 and 2,7000 BCE, also called Hafit tombs after Jebel Hafeet, a mountain in the north of Oman. There are more than 100,000 scattered across the country. Many of these tower-shaped structures have been well preserved. The Kubaykeb tombs near Al Jayla village at the top of Jabal Bani Jabir are 8 m high and stand 1,800 metres above sea level (Fig. 6.28). Similar tombs cluster on a ridge near Wadi Al Ayn village close to Jabal Al Misht. Some of these tombs were inscribed on the World Heritage List in 1988. In general, they are dry stone constructions consisting of circular towers made of slabs of stone without mortar. They have a diameter at the base of 4–8 m, tapering after a height of 6–8 m to a flat top. The inside is usually corbelled and above a relatively small aperture probably used as a collective burial chamber, the structure is often solid. The entrance is at crawl height and directed towards the east. They are seldom found in an oval or rectangular shape. Robust and dominant, these towers were built to last and built to be visible. Despite being vandalized repeatedly over thousands of years, they are still standing firm. They are scattered across the mountain tops above the small tribal oases that punctuate the dry wadis as if the ancient people wanted them to serve as a warning to strangers that their village was protected by the dead (Fig. 6.29).

From 2,700 BCE until the end of the Bronze Age in 2,000 BCE, a new type of circular tomb appeared known as Umm an-Nar tombs. They are also a circular, but not as tall as Hafit towers.

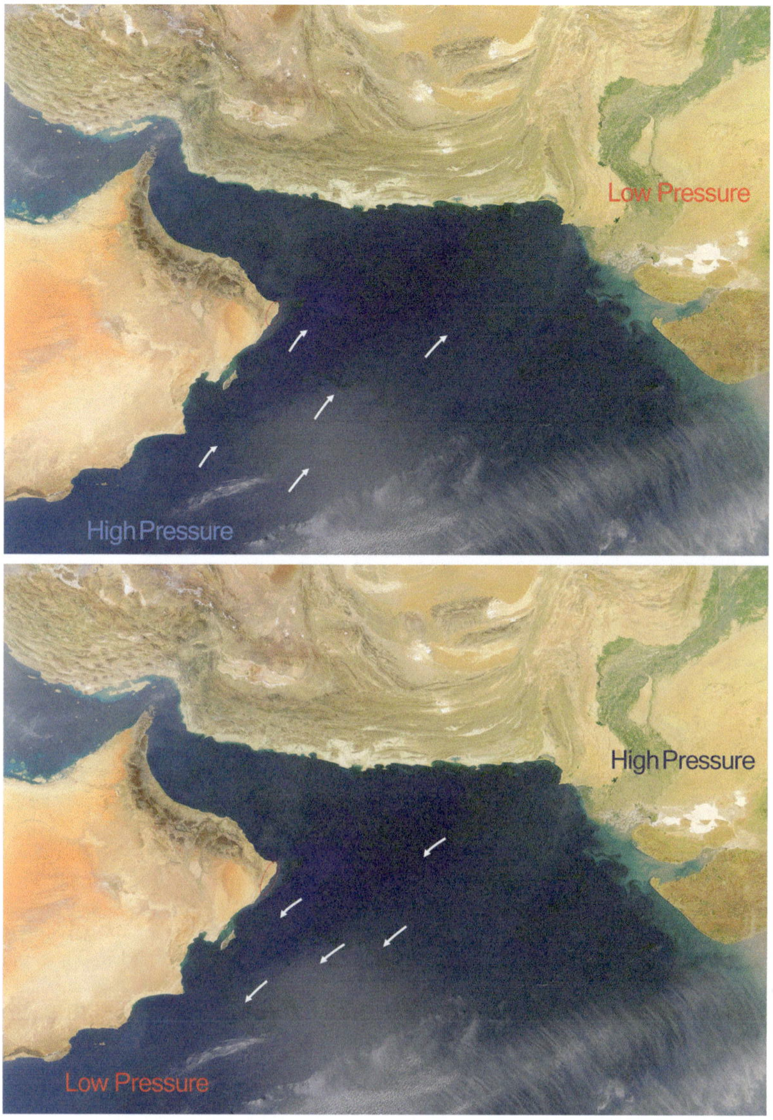

Fig. 6.25 Eastern coast of Oman is one of the most productive marine ecosystems anywhere on earth. The coast is heavily influenced by summer (the top figure) and winter (the bottom figure) monsoons (the satellite images are from NASA). These monsoons are primarily related to the larger thermal capacity of water compared to the land surfaces. During summer, the land surface heats more rapidly, creating a zone of low pressure which generates strong winds that blow from SW to NE over the Arabian Sea and the eastern coast of Oman. During winter, the cool Asian landmass creates a zone of high pressure compared to the ocean, which does not change temperature significantly between the seasons; therefore, the winds move from NE to SW. The strength of the wind is related to the temperature differential between water and land. The reversal of wind systems (trade winds) between summer and winter was recognized by mariners since the dawn of history. The red circle indicates the locations of Ras Al Had, Ras Al Jinz and Ras Al Khabbah, where ancient people chose to live since thousands of years ago. Recent climatic changes might lead to smaller variations in temperature between land and water, resulting in weakened wind systems

Fig. 6.26 Aerial photo is from Ras al Jinz and shows three archaeological sites, RAJ-2, RAJ-2 and RAJ-3 (Photo: Joint Al Had Project). The arrow represents the direction of the NE-directed summer monsoon and the strong coastal upwelling and easterly wind that results from this phenomena. This leads to the replacement of the nutrient-poor warmer surface water of the ocean by a nutrient-rich cooler water, which often results in phytoplankton blooms (see, for example Fig. 6.25), which in turn supports the growth of fish, marine mammals and seabirds. This phenomenon has supported fisheries for thousands of years and attracted humans to continuously occupy this part of Oman at least since the Holocene

They usually contain two doors on the outside. In general, Umm an-Nar tombs are more sophisticated in terms of design and more spacious and firmer at the bottom than Hafit tombs (Fig. 6.30). Their interior is also more complex and structured. They are usually scattered across the plains and small hills near the settlements instead of at the highest point on ridges or mountains. Perhaps they were intended to demarcate a particular community's territory. Thousands of these tombs have vanished from the landscape, their stones providing a convenient source of building materials for the construction of modern houses and the grave goods buried with the dead looted in ancient times. Perhaps their position on the plains is what attracted robbers and made it easier for people to reuse their stones unlike the mountain tombs of the Hafit Period. There are eleven Umm an-Nar Tombs in Bat village in Ibri (Fig. 6.31), and they can be found in Wadi Al Faj in wilayats Mahada and Yanqul and in Bisya village in Wilayat Bahla. They are also scattered across the eastern coast of Oman such as at Ras Al Jinz. They contain the remains of multiple burials as their size had increased over time. Grave goods include beads made of different materials in addition to beautifully engraved Sumerian pottery jars. Some of these were used to bury infants as it was the customs then to bury children in pottery jars outside the stone tombs. It

Fig. 6.27 Settlement of Ras Al Had (HD-6) is dated to 3,100 to 2,700 BCE. It includes many tools from the start of the Bronze Age. From left to right at the bottom, copper fish hooks, a net sinker and fish remain. (Photos: Joint Al Had Project.)

is noteworthy that these archaeological sites have been used repeatedly through the ages making the dating of artefacts problematic.

The communities of the Hafit and Umm an-Nar periods successfully exploited natural resources and were able to live in relatively dry locations by collecting rainwater or digging primitive water wells in wadi beds. They created beautiful oases with an effective agricultural system and sophisticated irrigation systems to cover large agricultural surfaces for the cultivation of palm trees, barley and other food crops. The domestication of livestock such as donkeys became widespread, and they used dogs for herding and guarding. The community developed complex rituals and a unique social system to bury the dead and demarcate their geographical territory. They possessed wealth and power and also began to develop technological and craft skills. This is the time when political, tribal and economic alliances began to prove significant in determining levels of power and ranges of influence. Moreover, the emergence of large oases marked the rise of communities endowed with a highly influential centralized authority (Fig. 6.32).

In the middle of the 3rd millennium BCE, the name of Majan appeared for the first time in the Sumerian cuneiform texts of the Mesopotamian Empire. Oman had by then become an integral part of a great Arab civilization, and the many magnificent oases with their surrounding mountains had become a source of important commodities within the scope of regional and global trade, most notably copper, diorite and marine products. Majan was also a base for the re-export of timber imported from India. Dozens of ancient mining sites have been found in Oman. They contain copper ores with different compositions which can be identified through the slag left behind during the smelting process. The amount suggests that there was a substantial trade in copper with much being exported abroad as well as being used domestically. Large settlements

Fig. 6.28 Hafit tombs in Al Jaylah (above) and Wadi Al Ayn (below)

were created near the copper extraction and smelting sites. They comprise a number of homes and huge furnaces for copper smelting. They also include mills and workshops for the production of bronze urns and tools. We can imagine what happened then being similar to what is happening today in terms of the extraction of crude oil and refining with a small part for local consumption and the remainder for export abroad.

With the emergence of large oases and Oman having gained a presence on the trade map with India, Persia and Mesopotamia, a network of commercial shipping lines began to appear across the Arabian Sea, the Sea of Oman and the Arabian Gulf. Over time, Omanis became pioneers in sailing and navigation and Ras Al Had, Muscat, Sohar, Musandam and Dhofar ports became international hubs for the transport of goods. Internally, a network of trade routes was established for the transport of goods from the interior to the coast. Oman began to be in close contact with its neighbours. The most famous Omani artefact to appear in this period is the statue of Gudea, the ruler (ensi) of the state of Lagash in Southern Mesopotamia, which dates back to 2,200 BCE. The statue was apparently made of Gabbro rock from the Oman Ophiolite layer. In return, there was a flow of different

Fig. 6.29 A number remain from the early Bronze Age. The top pictures show two typical Jemdat Nasr jars that were imported from Mesopotamia and found in various tombs in Oman. The jars are about 10 cm long each. On the left, an imprint of a sorghum spikelet cast in a mud brick (photo: Lorenzo Costantini). Below are two teeth of animals from HD-6, on the left possibly for a donkey and on the right for a camel (Curci, 2007), dated between the end of the 4th and the beginning of the 3rd millennium BCE. They could indicate a domestic use for transport

Fig. 6.30 Graves from different periods in Ras Al Had indicate continuous occupation (Munoz et al., 2008)

Fig. 6.31 A large mid-3rd millenium BCE tomb from the Umm an-Nar Period in Bat. They contain 'sugar-lump' stones that were brought from nearby outcrops. The photo on the left shows the subterranean chambers of a similar tomb in Ras al jinz (RAJ-1), from the mid-3rd millenium BCE. The chambers normally contain remains of many individuals buried together (photo: Joint Al Had Project)

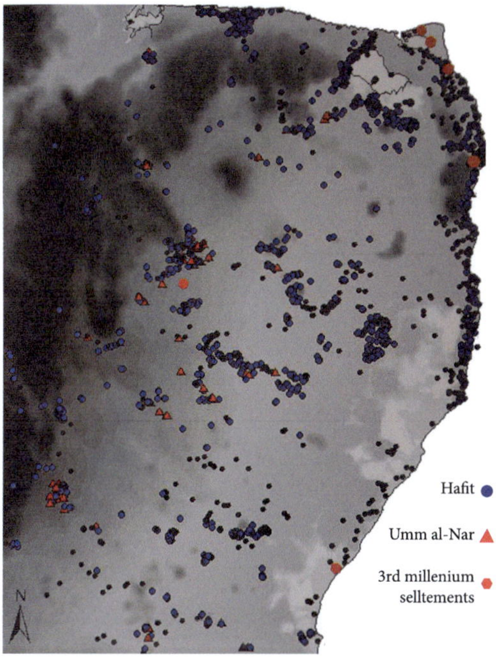

Fig. 6.32 A map showing the locations of the Hafit, Umm an-Nar and the 3rd millenium settlements around Ras Al Had and Ras Al Jinz. It shows how densely populated the area was (Map: Joint Al Had Project/Jessica Giraud)

commodities coming from India and Mesopotamia to Oman, such as timber, pieces of pottery, bitumen used in shipbuilding and metals like gold. Domestic animals such as chickens came from abroad at that time, too. At the end of the 3rd millennium BCE and the beginning of the 2nd millennium BCE, the Omani frankincense trade was established. During the same period, impressive stone seals and beautiful stone engravings appeared. The most famous of these was found in Hasat bin Salt in Wilayat Al Hamra, which depicts a group of men with a woman skilfully and exquisitely carved. Perhaps this stone carving represented some kind of religious ritual that used to take place on the slopes of towering mountains.

With the beginning of the 2nd millennium BCE, the area was undergoing another transitional stage marked by the emergence of new types of tools, metal pots and pottery vessels. The dead were buried individually in underground stone tombs not collectively as in the past (Fig. 6.33). These tombs were usually smaller and less well organized compared to Umm an-Nar types. This period was one of decline. Fewer buildings were constructed and the population shrank. It remained a puzzle for archaeologists known as the 'Wadi Souq Period' (from 2,000 to 1,300 BCE). The name refers to Wadi Souq near Sohar, where, along with Wadi Jizzi, artefacts of the period were unearthed for the first time in the 1970s. For many, this period represents a dark age of wars that broke out after the arrival of invaders from outside Oman and the commencement of fights over resources. It was also characterized by the spreading of economic crises and the decline of trade with Mesopotamia. As a result, some oases were abandoned, the nomadic lifestyle began to prevail, and camels were widely domesticated. There were fewer Wadi Souq-type settlements—far fewer than in the Hafit and Umm an-Nar periods—and they are scattered over a far wider geographical area, indicating a decline in populations. This may be also linked to a Bedouin lifestyle, which does not leave behind it any artefacts that may be preserved for thousands of years. The largest settlements of this period are found in Bat and Bawsher, Samad Ash Shan and Wasit in addition to Wadi Al Jizi. However, the oldest of these existing settlements is located in Ras Al Jinz where men lived on higher ground, some 20–30 m above sea level after the migration of the early Bronze Age settlements at the bottom (refer to Fig. 6.26). Perhaps by moving to live on top of the headland at Ras Al Jinz, the people were protecting themselves against the conflicts prevailing at that time. This preference for settling in geographically fortified places probably explains the disappearance of archaeological sites in Masirah Island during the Hafit and Umm an-Nar periods. They were abundant during the earlier stone ages and only reappeared during the Wadi Souq Period (Fig. 6.34).

The lifestyle of the Wadi Souq period was, in many ways, very similar to that of previous times. Coastal dwellers lived on catching and drying fish, whereas the inhabitants of the oases and the desert lived on herding and agriculture.

Fig. 6.33 A single burial tomb from the Wadi Souq Period in the Wilayat of Bawshar with two smaller tombs on the side (the photo is from the Joint Arabian Excavations at Bawsher). These tombs are significantly different from the large tombs of the Hafit and Umm an-Nar periods because they are dug in the ground and their sides lined with stones

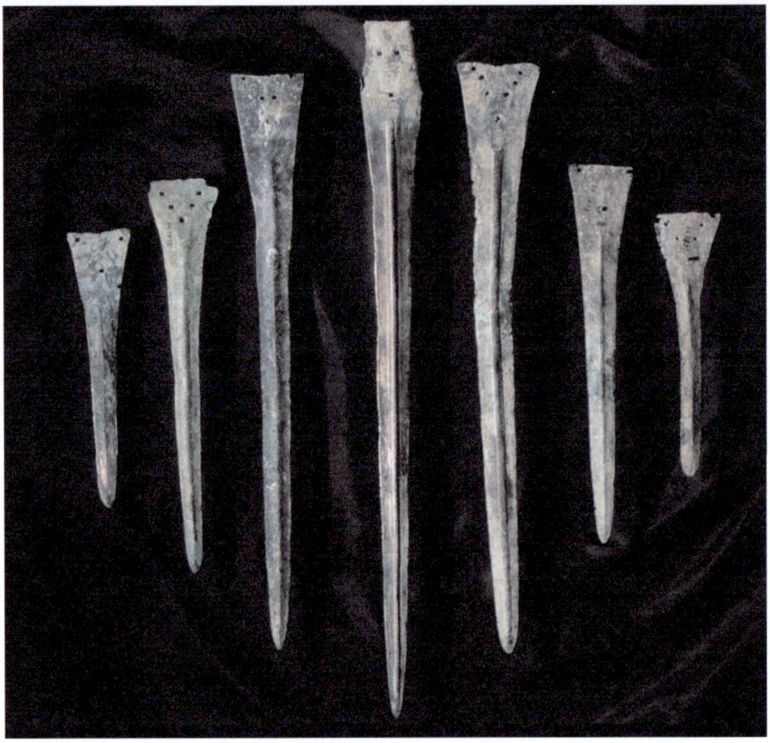

Fig. 6.34 A number of copper daggers found in a collective grave from the Wadi Souq Period at Al Wasit (photo by the German Mining Museum in Bochum)

However, the stone, metal and pottery tools became larger, more sophisticated and less perishable. Daggers, swords and spears made of Omani copper were widely used—perhaps a natural result of the absence of security, the growing importance of self-defence and the need for the acquisition of arms. Men chose to live in less complicated settlements so as to enable their community members to move and travel whenever the need arose. It seems that the general

Fig. 6.35 A burial from the Iron Age in the Wilayat of Bawshar. It consists an aggregation of individual tombs (photo is from the Department of Excavation and Archaeological Studies)

culture of the community was influenced by external interventions and repeated migrations (Fig. 6.35).

6.5.3 Iron Age

The Iron Age in Oman is divided into several periods and is a very special time in Oman's history. It is the age during which Oman established its cultural identity and affirmed its presence as a dynamic civilization: copper was extracted in large commercial quantities, further irrigation channels (aflaj) were built, sea navigation became more sophisticated, frankincense became a widely traded commodity, and agriculture was largely developed. Despite being called the Iron Age, iron was not widely used in Oman until the eighteenth century BCE. The First Iron Age Period (Iron Age I) falls between 1,300 and 1,100 BCE and follows the Wadi Souq Period, but artefacts from this time are rare. Few have been unearthed, with the exception of a particularly rich find at Jabal Al Hawra in Wilayat Nizwa. Among the contents of a tomb were a large set of spears, daggers and copper rings in addition to many pieces of pottery (Fig. 6.36). Archaeologists hold that the tomb might have belonged to a rich warrior who lived in a fortified settlement in Nizwa at the end of the 2nd millennium BCE (during the First Iron Age Period). As a result, this time is often referred to as the Nizwa Period.

The Second Iron Age Period spans the years 1,100 and 600 BCE. It is known as the Lizq Period after the village of the same name in Wilayat of Al Mudaybi. The archaeological sites of this period are common across many regions in Oman including Al Muyassar town in Wilayat of Al Mudaybi (Fig. 6.37) and Al Raki in Wilayat Yanqul. They are found in the mountains, wadis and coasts and, unlike the Nizwa

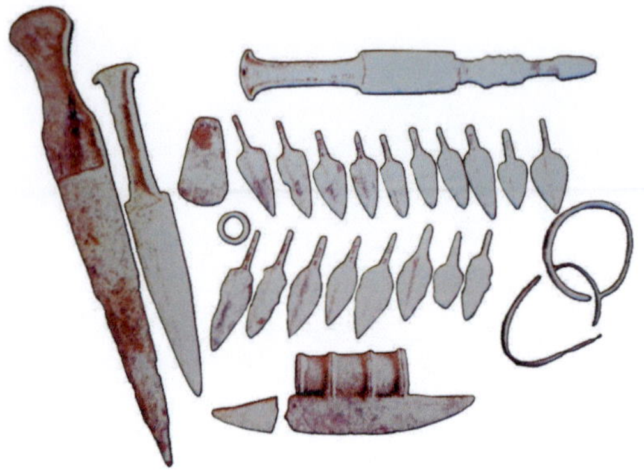

Fig. 6.36 A set of tools from the Iron Age found in the 'warrior's grave' in Nizwa. (Photo: German Mining Museum, Bochum)

Period, provide solid evidence in terms of artefacts and remains. The villages of the Lizq Period are characterized by their solid buildings which usually comprise an outer wall with a single entrance with a group of houses inside, each containing a few rooms. These houses are separated by relatively wide spaces. As for funerary monuments, most of them are individual 'beehive' tombs that come in circular or oval shapes. They contained a large quantity of spears, daggers, vessels and bracelets made of copper and bronze. Given the richness of grave goods, many of these tombs were robbed in antiquity, just like earlier Stone Age and Bronze Age tombs. In addition to fortified villages, the Lizq Period was characterized by mountain forts that were perhaps used as dwellings for the ruling dynasties or places of refuge during conflict. It was in the Lizq Period that Oman regained its presence among the major civilizations of the world, a position it has retained to this day. Historical sources tell the story of the King of Qadē who lived in Izki and sent gifts to the Assyrian King in 640 BCE. There is some evidence that the Maka and Makkan mentioned in trilingual Achaemenid inscriptions is Magan (Oman) and the old Persian equivalent is Qadē. At Susa and Persepolis, carvings show Maciyā—the inhabitants of Maka—wearing traditional swords of the style found in Iron Age graves in Oman.

During the late Iron Age, man-made features known as triliths (Fig. 6.37) appeared in a landscape occupied from ancient times. They are formed of rows of three upright stones, sometimes topped with a fourth, flat stone, surrounded by a kerb and often infilled with gravel or pebbles, and can extend 25 m in length and occupy wadi terraces extending for several kilometres. In front of each trilith, there is often a firepit surrounded by four large boulders. Carbon 14 dating suggests they were in at the height of use around 300 BCE, but archaeologists suggest that the wadi terraces had been in continuous use since Bronze age times. Triliths spread north from Yemen, through Dhofar and along the coasts into the Sharqiyah region in the north of Oman.

The end of the Iron Age was marked by the Samad Period. The dates for this period are not well defined, but probably range from around 300 BCE–600 CE, representing the late Pre-Islamic Age in Oman. During that time, iron was widely used in the fabrication of tools, metal coins were struck, glassware was imported, new types of pottery vessels or colourful ceramics appeared, and camels were widely domesticated and became an integral part of people's life. In the locality of Samad, a camel carcass dating back to the late Iron Age was found—the oldest one discovered in the Arabian Peninsula so far. However, this was apparently the period during which the Persians occupied some of the major villages in Oman. Perhaps this explains the appearance of new types of ceramics found among the archaeological debris as well as the widespread use of permanent irrigation channels (Aflaj). Maybe, some of these tools and ideas were imported from Persia. Persian influence, however, was not to last, as Malik bin Faham expelled the Persians from Oman in the first centuries CE, returning the land to its rightful inhabitants (Fig. 6.38).

Fig. 6.37 Evidence of the Iron Age in Oman. Top: Iron Age steps to the fortress at Lizq. Bottom: Enigmatic Iron Age 'trilith' from Dhofar. Their purpose is unknown, but they are associated with fire pits and generally lie along wadis and possible trade routes (photos by Jill Newby)

Fig. 6.38 Typical Iron Age hill top fortress with surrounding dry stone walls (photo by Jill Newby)

6.5.4 Summary of Human's Life in Oman Since the Neolithic

Figure 6.39 summarizing the main characteristics of the archaeology of Oman since the Neolithic, mainly established from the graves of the Hafit, Umm an-Nar and Wadi Suq periods. Burial practices and the tools found in graves are revealing. During the Neolithic, the graves are primary and mostly individual. At the end of the 4th millennium, during the start of the Hafit Period, the graves became monumental objects, punctuating the landscape as markers of the territory. During the 3rd millennium, during the Umm an-Narr Period, graves became larger, circular and divided to many chambers which contained dozens or hundreds of individuals, reflecting larger communities and a growing population

The graves of the Hafit and Umm an-Narr periods are widespread across Oman in both coastal and interior regions. Major agricultural settlements appeared along the wadis and on the flanks of the northern and southern mountains of Oman. The establishment of falaj technology, which provides a reliable water supply across several kilometres within an arid climate, was certainly one of the main elements in support of the development of large oases and the initiation of farming communities in the interior of Oman. Along the coast, for example at Ras Al Had Omani, fishermen managed

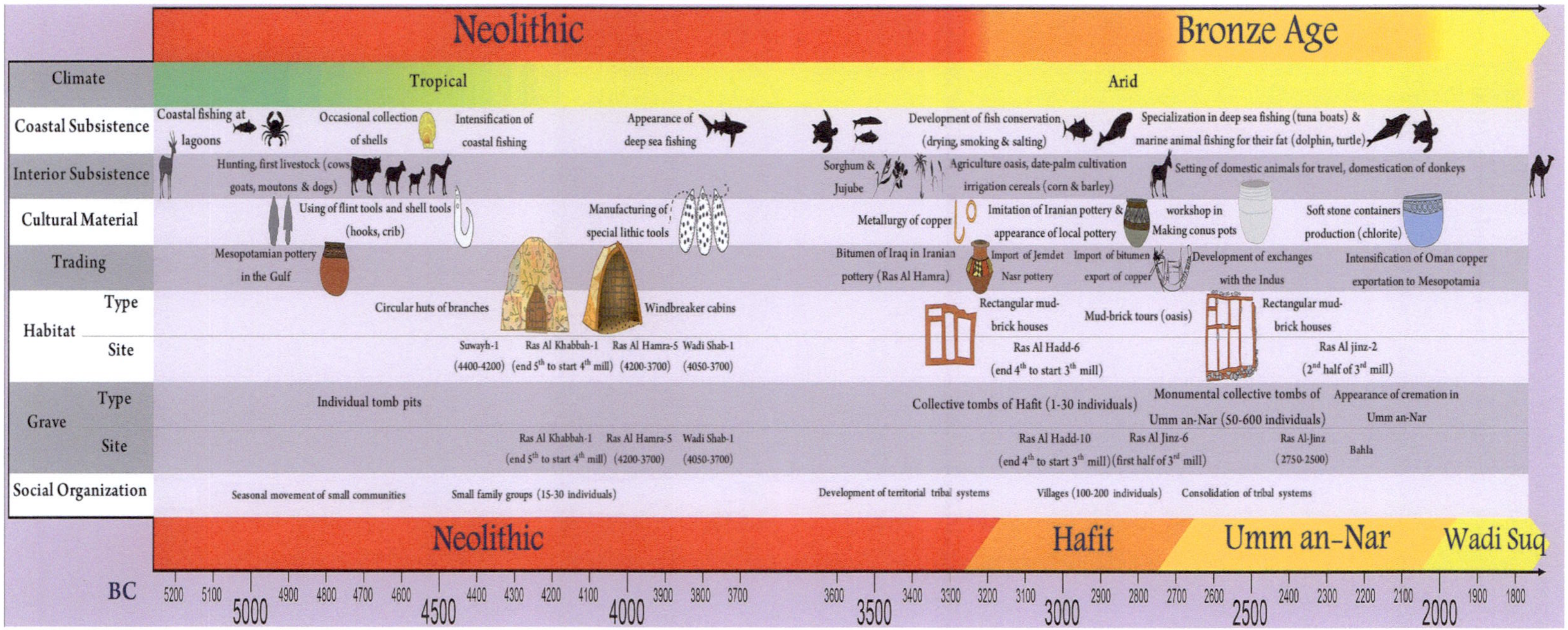

Fig. 6.39 A summary of the history of humans in Oman since the 6th millenium BCE. The figure is modified and translated from Munoz and Cleuziou, 2008

to establish trade routes across the Indian Ocean. Agricultural production, copper extraction and marine industries have supported a strong integration and exchange between the local oases and coastal communities and resulted in the appearance of regional trade routes.

About the Geological Society of Oman

The Geological Society of Oman (GSO) is one of the first vocational societies established in Oman. Since its foundation in 2001, the GSO has become one of the leading societies for the preservation of our geological heritage. It provides seminars, lectures, workshops and field trips in various geological fields. Currently, it is involved in designing and establishing a number of geological museums in Oman. There are more than 1,000 members including well-known geological experts. The society also provides geological consultancy and assistance to various ministries in Oman.

M. H. Al Kindi, *Evolution of Land and Life in Oman: An 800 Million Year Story*,
https://doi.org/10.1007/978-3-319-60152-6

A Map of the Locations Mentioned in the Text

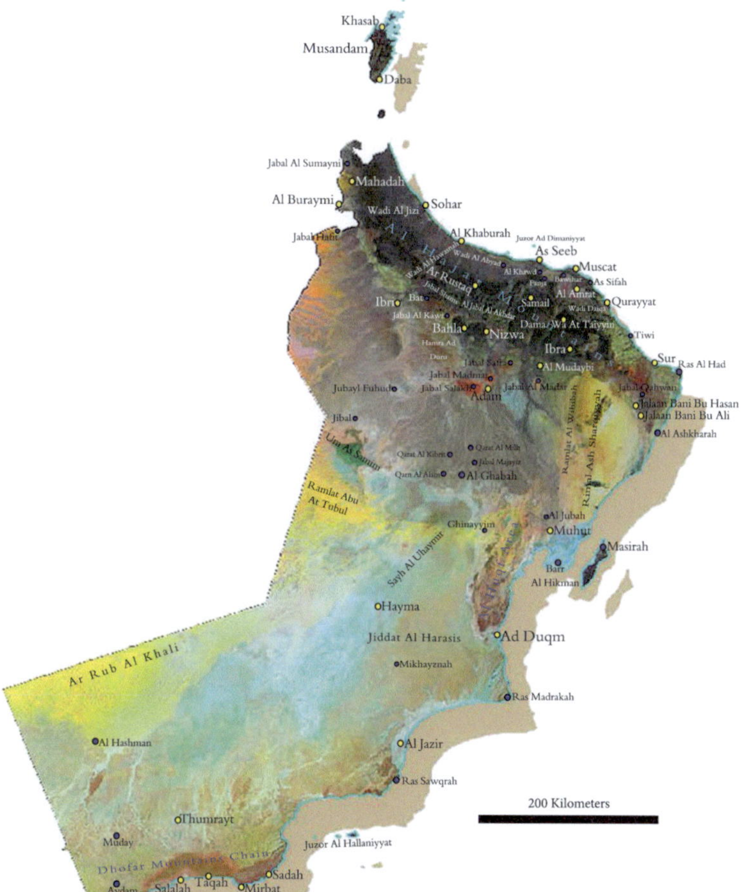

Names as per the National Survey Authority, Sultanate of Oman, March 2014

Archaeological places are shown in figure 199.

M. H. Al Kindi, *Evolution of Land and Life in Oman: An 800 Million Year Story*,
https://doi.org/10.1007/978-3-319-60152-6

Glossary

Accretion Gluing together. In astrophysics, planets are formed by accretion when gravity pulls in matter from space. In geology, accretion occurs when a land mass grows, either through volcanic action or plate tectonics.

Age The smallest subdivision of geological time. There are several ages in an epoch. See 'eon'.

Amphibia Animals born in water, but which breathe air when they grow. Frogs are amphibians.

Arthropods Animals with an exoskeleton like lobsters and scorpions.

Asphyxiation Death through lack of oxygen or choking.

Biota All cellular life comes under the super-domain name of 'biota'.

Carnivorous Meat-eating. Leopards and wolves are carnivores.

Cephalopod Class of animals with symmetrical bodies and arms or tentacles. Squid and octopus are cephalopods.

Chert Hard, brittle rock often used by ancient people for making stone tools.

Clast Fragment of geological rubbish! Fragments of bigger rocks are clast.

Conglomerate Rock formed of clasts stuck together by smaller fragments of rock to form new layers.

Dike A sheet of rock which forms in a crack between preexisting rocks. They may be sedimentary or igneous.

DNA Deoxyribonucleic acid, one of the essential ingredients for life.

Eon The largest unit of geological time. If an eon is like a century, an era is like a decade, a period like a year, an epoch like a month and an age like a week.

Epoch The third subdivision of geological time. There are several ages in an epoch, and there may be several epochs in a period. See 'eon'.

Era The first subdivision of geological time. There may be a number of eras in an eon, and an era may be divided into several periods. See 'eon'.

Eukaryote Early life form: cells contain a nucleus.

Exoskeleton A skeleton on the outside of a creature. Scorpions have exoskeletons.

Gastroliths Stones eaten by some animals to help grind up food in the stomach.

Herbivorous Plant-eating. Camels and oryx are herbivores.

Igneous rock Rock formed as magma or lava cools. Basalt is an igneous rock.

Invertebrate Life forms with no spinal column or backbone. Corals and worms are invertebrates.

M. H. Al Kindi, *Evolution of Land and Life in Oman: An 800 Million Year Story*,
https://doi.org/10.1007/978-3-319-60152-6

Lithostratigraphic unit Rock unit, a deposit with similar characteristics, such as a layer of ash deposited after a volcanic eruption.

Magma Molten rock, often extruded during volcanic eruptions.

Mammal An animal which bears live young. Elephants, whales and humans are mammals.

Metamorphic rock Rock formed from other rock through heat or pressure. Marble is a metamorphic rock.

Methanotrophs Prokaryotes which use methane for food.

Mitochondrion The power plant of a cell, generating energy for an organism from a eukaryote to a human.

Obduction The movement of oceanic plates over the top of continental plates. The pushing down of the underlying rock is called subduction.

Omnivorous Eating anything. Seagulls and (most) humans are omnivores.

Ophiolite Rock which was once part of the earth's oceanic crust. Its name comes from the Greek words ophis meaning 'snake' and lithos meaning 'stone' because many Ophiolites are the mottled green colours of snakes.

Period The second subdivision of geological time. There are a number of periods in an era, and a number of epochs in a period. See 'eon'.

Petrified It comes from the Greek word petro meaning 'stone'. It is used to refer to trees which have fossilized or turned to stone.

Photosynthesis The process of converting sunlight to energy. Organisms containing the green substance chlorophyll (like the leaves on date trees) use photosynthesis.

Phylum A taxonomic rank below kingdom and above class.

Plankton A broad group of organisms which live in the sea, but cannot swim against the current. They are a major food source for many marine species, like jellyfish and whales.

Prokaryote The first life form: single-celled creatures without a real nucleus.

Rheology The study of the flow of matter, such as the movement of tectonic plates or mud flows.

RNA Ribonucleic acid, one of the essential ingredients for life.

Sabkha Salt flat

Sedimentary rock Rock formed by deposition of materials in layers. Sandstone is a sedimentary rock.

Stratigraphy A branch of geology which studies rock layers (strata) and layering (stratification). It is primarily used in the study of sedimentary and layered volcanic rocks.

Suture zone The area where one tectonic plate is accreted to another.

Teleost Ray-finned fish. There are over 26,000 kinds of fish alive today—most are teleosts.

Viscosity A measure of the resistance of a liquid. Honey and magma are viscous, petrol and water are not.

References

In the Shadow of the Ancestors, the Prehistoric Foundation of the Early Arabian Civilization in Oman, Serge Cleuziou & Maurizio Tosi, Ministry of Heritage and Culture.

Fundamentals of Historical Geology, Prof. Abdul Jalil Huwaidi and Prof. Mohammed Ahmed Heikal, Faculty of Science, Al-Azhar University, Al Dar Al Arabiya for Books, 2004.

Oman, Faces and Places, Articles from PDO News Magazine, 2008.

The Concise Geology of Oman, by: Prof. Ken Glennie, edited and translated by: Mohammed Al Kindi and Mahmoud Mahrooqi, produced by: the Geological Society of Oman, 2012.

Field Guide to the Geology of Oman, by Samir Hanna, published by the Historical Association of Oman, 1996.

Oman's Geological Heritage, By Mike Hughes Clarke, PDO Special Publication, 2nd Edition, Edited by Ken Glennie, 2006.

The Evolution of Human Populations in Arabia, edited by Michael D. Petraglia and Jeffrey Rose, Springer, 2009.

Amthor, J., Grotzinger, J., Schro der, S., Bowring, S., Ramezani, J., Martin, M. Matter, A.(2003) "Extinction of Cloudina and Namacalathus at the Precambrian-Cambrian boundary in Oman". Geological Society of America, v. 31, no. 5: 431–434.

Anne S. Schulp, Matteo Belvedere, Sobhi Nasir & Mark Wheeler (2011): An Artiodactyl Tracksite at Musandam Peninsula, Sultanate of Oman, Ichnos, 18:4, 192-196.

Al-Ghammari, M., G.A. Booth and F. Paris 2010. New chitinozoan species from the Saih Nihayda Formation, Middle Ordovician of the Sultanate of Oman. Review of Paleobotany and Palynology, v. 158, p. 250–261.

Curci, A, Carletti, M, Tosi, M (2014) The camel remains from site HD-6 (Ra's al-Hadd, Sultanate of Oman): An opportunity for a critical review of dromedary findings in eastern Arabia. Anthropozoolgica 49(2): 207–222.

Crasquin-Soleau, S. 2003. New ostracod (Crustacea) species (Geffenina wangi) of the Middle Permian from the Jebel Gharif (Central Oman). Journal of Micropaleontology, v. 22, p. 101–105.

Dovan, S., Miller, G., Sansom, I., Hewards, A. & Schreurs, J. (2011) "A Laurentian Iocrinus Hall (crinoidea, disparida) in the Dapingian or Darriwilian (middle Ordovician, Arenig) of Oman" Palaeontology, Vol. 54, Part 3:.525–533.

Forke, H. C. (2012) "Fossil Guide to Khuff Outcrop equivalent: Saiq and Mahil Formations Al Jabal Al-Akhdar, Saiq Plateau, Sultanate of Oman" PDO's Internal Report.

Fortey R. A. & Heward A. P. (2015). A new, morphologically diverse Permian trilobite fauna from Oman. Acta Palaeontol Pol., 60:(1): 201–216.

Gheerbrant, E., Thomas, H., Sen, S., & Al Sulaimani, Z. (1995). New Oligopithecinae Primate(Simiiformes) From The Early Oligocene Of Taqah, Sultanate-Of-Oman. Comptes Rendus De L Academie Des Sciences Serie Ii Fascicule A-Sciences De La Terre Et Des Planetes, 321(5), 425–432.

Grotzinger, J.P., Watters, W.A., Knoll, A.H. (2000). «Calcified metazoans in thrombolitestromatolitereefs of the terminal Proterozoic Nama Group, Namibia». Paleobiology 26 (3):334–359.

Hahn, G., and H. D. Pflug (1985). «Die Cloudinidae n. fam., Kalk-Röhren aus dem Vendium und Unter-Kambrium». Senckenbergiana lethaea 65: 413–431.

Hansen, J., M. Sato., P. Kharecha., D. Beerling., R. Berner., V. Masson-Delmotte., M. Pagani., M. Raymo., D. L. Royer., and J. C. Zachos (2008). Target atmospheric CO2: Where should humanity aim? J Sci Atmos Open Atmos 2: 217–231.

Isozaki, Y. & Aljinovic, D. (2009) "End-Guadalupian extinction of the Permian gigantic bivalve Alatoconchidae: End of gigantism in tropical seas by cooling", Palaeogeography, Palaeoclimatology, Palaeoecology 284: 11–21.

James, Z., Pagani, M., Sloan, L., Thomas, E. & Billups, K. (2001). «Trends, Rhythms, and Aberrations in Global Climate 65 Ma to Present». Science 292 (5517): 686–693.

Kozur H. W. & Wardlaw B. R. 2010. — The Guadalupian conodont fauna of Rustaq and Wadi Wasit, Oman and a West Texas connection. Micropaleontology 56 (1–2): 213–231.

Love, G., Grosjean, E., Stalvies, C., Fike, D., Grotzinger, J., Bradley, A., Kelly, A., Maya Bhatia, William Meredith, Colin E. Snape, Samuel A. Bowring,

M. H. Al Kindi, *Evolution of Land and Life in Oman: An 800 Million Year Story*,
https://doi.org/10.1007/978-3-319-60152-6

Daniel J. Condon & Roger E. Summons (2009) "Fossil steroids record the appearance of Demospongiae during theCryogenian period". Nature 457, 718–721.

Maryland Geological Survey E.S. No. 4, p. 38, Fig. 35, Crinoids Reconstruction.

Munoz O, Zazzo A, Bortolini E, Seguin G, Saliège J-F, Cleuziou S. 2008. Reconstructing the diet of the ancient fishermen of Ra's al-Hadd and Ra's al-Jinz (Sultanate of Oman) using radiocarbon dates. In: Les Déserts d'Afrique et d'Arabie: environnement, climat et impact sur les populations. Colloque de l'Académie des sciences, Institut de France, 8–9 September, Paris.

Munoz O. and Cleuziou S. (2008) – La tombe 1 de Ra's al-Jinz RJ-1 : une approche de la complexité des pratiques funé¬raires dans la péninsule d'Oman à l'âge du Bronze ancien, in J. M. Cordoba, M. Molist, M. C. Perez, I. Rubio et S. Marti¬nez (éd.), Proceedings of the 5th International Congress on the Archaeology of the Ancient Near East (ICAANE), actes du colloque international (Madrid, 3-8 avril 2006), Madrid, Universidad Autónoma de Madrid, p. 627–644.

Nelson, S.J. 1977. Mississippian syringoporid corals, southern Canadian Rocky Mountains. Bulletin of Canadian Petroleum Geology 25: 518–581.

Oekentorp, K., Montenat, C. and Fontaine, H. (1978): Eine kleine Korallenfauna aus dem unteren Oberperm von Saiq. Oman (Arabische Halbinsel).—N. Jb. Geol. Paläont. Abh.,155/3, 374–397, 7 Figs., Stuttgart.

Orchard M J. Taxonomy and correlation of Lower Triassic (Spathian) segminate conodonts from Oman and revision of some species of Neospathodus. J Paleont, 1995, 69: 110–122.

Penney, R.A. and P.L. Osterloff 2002. An improved palynostratigraphic zonation for the glacigenic Al-Khlata Formation: refinements and applications. 5th Middle East Geosciences Conference, GEO 2002. GeoArabia, Abstract, v. 7, no. 2, p. 285–286.

Raup, D. & Sepkoski, J. (1982). «Mass extinctions in the marine fossil record». Science 215: 1501–1503.

Rickards, R. B. (2007) "Ordovician Graptoloids from Oman" internal PDO report.

Rieu. R., Allen, P., Cozzi, A., Kosler, J., & Bussy, F. (2007) "A composite stratigraphy for the Neoproterozoic Huqf Supergroup of Oman: integrating new litho-, chemo- and chronostratigraphic data of the Mirbat area, southern Oman". Journal of the Geological Society, London, Vol. 164: 997–1009.

Romine & Stuart-Smith (2003) "Regional Seismic Interpretation: Western Deformation Front, South Oman Salt Basin", PDO Internal Report, used as a basis to capture paleogeographic maps.

Rose. J. I., & Usik, V. I. (2009) The "Upper Paleolithic" of South Arabia. In: Petraglia MD, Rose JI, editors. The Evolution of Human Populations in Arabia: Paleoenvironments, Prehistory and Genetics. Netherlands: Springer Academic Publishers. pp. 169–185.

Sarnthein, M. 1972. Sediments and history of the postglacial transgression in the Persian Gulf and Northwest Gulf of Oman. Marine Geology, 12: 245–266.

Sansom, I., Miller, G., Heward, A., Davies, N., Booth, G., Fortey, R. & Paris, F. (2009) "Ordovician Fish from the Arabian Peninsula" Palaeontology, Vol. 52, Part 2: 337–342

Seibold, E., &Vollbrecht, K.: Die Bodengestalt des Persischen Golfs. – "Meteor" Forschungserg., C2, 29–56, Stuttgart 1969.

Selden PA, Huys R, Stephenson MH, Heward AP, Taylor PN. Crustaceans from bitumen clast in Carboniferous glacial diamictite extend fossil record of copepods, Nat Commun. 2010, vol. 1 pg. 50

Smith, R. (2010) "A large ammonite from the Natih Formation" Al Hajar Newspaper by the Geological Society of Oman, Issue 17: 17.

Thomas H, Roger J, Sen S, Pickford M, Gheerbrant E, Al-Sulaimani Z, Al-Busaidi S. (1999) "Oligocene and Miocene terrestrial vertebrates in the southern Arabian Peninsula (Sultanate of Oman) and their geodynamic and palaeogeographic settings" In: Andrews PJ, Hill A, Whybrow P, editors. Fossil vertebrates of Arabia. New Haven: Yale University Press: 430–442.

Wellman, C., Osterloff, P. * Mohiuddin, U. (2003) "Fragments of the earliest land plants" Nature, Vol. 425: 282285.

Wikipedia, the free encyclopedia: http://en.wikipedia.org.

Xiong, J. & Bauer, C. (2002) "Complex Evolution OF Photosynthesis". Annual Review of Plant Biology, Vol. 53: 503–521.

MIX
Papier aus verantwortungsvollen Quellen
Paper from responsible sources
FSC® C105338

If you have any concerns about our products,
you can contact us on
ProductSafety@springernature.com

In case Publisher is established outside the EU,
the EU authorized representative is:
Springer Nature Customer Service Center GmbH
Europaplatz 3, 69115 Heidelberg, Germany

Printed by Libri Plureos GmbH
in Hamburg, Germany